ADDED TO THE
BOSTON LIBRARY
20th day of December 1870
to be returned in seven days.

A fine of three cents will be incurred for each day
this volume is detained beyond that time.

BOSTON
MEDICAL LIBRARY
8 THE FENWAY

# THE SPAS

OF

## BELGIUM, GERMANY, SWITZERLAND, FRANCE, AND ITALY,

A

HAND-BOOK OF THE PRINCIPAL WATERING PLACES ON THE CONTINENT.

DESCRIPTIVE OF THEIR NATURE AND USES IN THE TREATMENT OF CHRONIC DISEASES, ESPECIALLY GOUT, RHEUMATISM, AND DYSPEPSIA; WITH NOTICES OF SPA LIFE, AND INCIDENTS IN TRAVEL.

BY

THOMAS MORE MADDEN, M.R.I.A.

LICENTIATE OF THE KING'S AND QUEEN'S COLLEGE OF PHYSICIANS IN IRELAND; MEMBER OF THE ROYAL COLLEGE OF SURGEONS, ENGLAND; AUTHOR OF "CHANGE OF CLIMATE IN PURSUIT OF HEALTH," "OBSERVATIONS ON INSANITY AND CRIMINAL RESPONSIBILITY,"
ETC.

London:

T. C. NEWBY, 30, WELBECK STREET,

1867.

ALL RIGHTS RESERVED.

1 F₂ 377.

DEDICATION TO

# WILLIAM BEATTIE, Esq., M.D.,

FELLOW OF THE ROYAL COLLEGE OF PHYSICIANS, LONDON,

FORMERLY PHYSICIAN EXTRAORDINARY TO H. M. WILLIAM IV.

ETC., ETC., ETC.

---

My Dear Sir,

Amongst the members of the Medical profession to whom a work of this kind could be appropriately inscribed, there are none to whom it could be so fitly dedicated as yourself, distinguished as you are, no less in Medical than in Miscellaneous Literature of the highest order, cultivated with a view to the best interests and the noblest ends of those pursuits which you adorn.

I am actuated, however, by other motives for connecting this work of mine with a name so honoured and respected as yours—motives founded on that long tried and steadfast friendship, which neither time nor distance has ever interrupted between you and those with whom I am most nearly and dearly connected: and I may add also on that uniform kindness of yours towards myself, which I have had gratifying proofs of on so many occasions.

I am, my dear Sir,
Yours very faithfully,
THOMAS MORE MADDEN.

9, Great Denmark Street, Dublin,
1867.

# CONTENTS.

## INTRODUCTION.

### ON THE NATURE AND USES OF MINERAL AND THERMAL WATERS.

What Constitutes a Mineral Spring—Cold Mineral Waters—Their Origin—Carbonic Acid Gas in 'them—Its Influence in Mineral Springs—Thermal Sources—Their Mode of Formation in the Earth—Their Connection with Volcanic Action—Classification of Mineral Springs—The Muriated, Saline, Acidulous, Alkaline, Chalybeate, Iodated, and Bitter Waters of Europe—Their Nature and Effects—Various Theories—Author's Opinions—Influence of Change of Climate, Scene, and mode of Life .................................................. 1

## CHAPTER I.

### DYSPEPSIA AND THE SPAS.

Nature and symptoms of indigestion—Moral and physical importance of a good digestion.—DIETETICS: A mixed diet best—Errors of ordinary habits of living—Abstinence—Rule for dyspeptics—Use and abuse of wine and stimulants—Treatment of indigestion—Benefits of travelling—Mineral waters employed in dyspepsia and hypochondriasis....... 30

## CHAPTER II.

### ON GOUT AND ITS TREATMENT BY MINERAL WATERS.

Preliminary remarks—Gout not confined to the fashionable world—Causes of the disease—Premonitory symptoms—Description of a fit of podagra—Letter of a gouty wit—Irregular gout and the gouty diathesis—Treatment—Opinions of the ancients—Author's method—Springs resorted to—Simple saline waters, Wiesbaden and Homburg—Saline Alkaline waters, Carlsbad and the Bohemian Springs—Carbonated Spas, Vichy, Fachingen, and Bilin—The English gout-curing watering places—General rules for gouty patients. ........................ 39

## CHAPTER III.

### LIFE AT THE GERMAN SPAS.

Comparison of German Spas with English Watering-places—Modes of Travelling in Germany—The Railways-Hotels—Dread of fresh Air and Water—Remarks on daily Cold Baths—Heterodox Notion —German Living. Table d'hôte Dinners—Rhine Wines—Voracious Appetite of Natives—How to diagnose a German—The Cursaals— Gambling; its universal Prevalence—Natural Divisions of German Society—Effects of Gambling—Its Physiological and Pathological Results .................................................................................................. 53

## CHAPTER IV.

### THE ART OF TRAVEL.

Its importance—English travellers—National love of travel—Its pleasures and advantages—Excuses for it—Classes of tourists—Hint to travellers—A scene at Spa—Comparisons odious abroad—Respect due to observances of others—Prejudices and luggage to be got rid of— Pedestrianism and travelling "en grand seigneur" compared ............ 70

## CHAPTER V.

### THROUGH BELGIUM TO CHAUDFONTAINE.

The voyage — Ostend — A metamorphosis — Effect of the sea on French and English women—Reminiscences of Belgium.—CHAUDFONTAINE: Description of the village—Its thermal sources—The baths and the author's experience of them—Their medicinal application ... 82

## CHAPTER VI.

### SPA.

Situation of the town—Its chief features—Population fixed and transitory—Doctors—Beauty of surrounding scenery—Advantages for valetudinarians—Hotels and living—The "Redoute"—The season— Characteristics of society here contrasted with other watering places— Analysis of the mineral springs—Detailed account of each source— Action of the waters—Patients who should be sent there..................... 91

## CHAPTER VII.

### AIX-LA-CHAPELLE.

The journey from Spa—A breakdown, of which the author avails himself to digress—Arrival in Aix—The Dragon d'Or—Dr. Velten—Account of AACHEN—The mineral springs—Analysis—Their common origin proved—Their physiological effects and remedial power—Influence on chronic rheumatism, gout, and cutaneous affections—The Baths—Their modes of administration and action—BORCETTE: Its hot springs and their uses.................................................................................. 105

## CHAPTER VIII.

### AIX-LA-CHAPELLE TO EMS.

Journey to Cologne—A house on fire—Cologne—The Rhine—Oberlahnstein—The village wedding and our luggage—The valley of the Lahn—EMS: Description of the town—The thermal sources—Their composition and effects—Diseases for which applicable—Mineral waters generally useless in cases of consumption—General remarks on this Spa 118

## CHAPTER IX.

### SCHWALBACH AND SCHLANGENBAD.

Departure from Ems—Eltville—The Gasthoff—Moonlight scene on the Rhine—Drive through the Rheingau—Our companions—SCHWALBACH:—The town—Hotels—Resources for invalids—Account of each mineral source—The baths and the Empress of the French—Cudgeling *versus* cajoling—Analysis of the springs—Their physiological, pathological, and curative effects—Diseases in which they should be used—Account of SCHLANGENBAD—The waters and baths and their uses.................................................................................................. 126

## CHAPTER X.

### WIESBADEN.

Arrival in Wiesbaden—A cheerful prospect—A cab stand *versus* an hotel—"Unprotected females"—Description of WIESBADEN—The Kochbrunnen—Analysis of the springs—Mode of using the water—Its effects—Diseases in which it is prescribed—The hot baths and their sanative influence—Cases in which this Spa should be interdicted .................. 146

## CHAPTER XI.

### WEILBACH AND SODEN.

The Journey—Orchards of Nassau—Traits of the Peasantry—WEILBACH—A High-Dutch Amazon—The Mineral Springs and Baths—Monotony of Spa-life here—Cases in which the Weilbach Water is employed—Route to Soden—Harvest Scene—A Contrast—Account of SODEN—Great Variety of Mineral Springs—Their Composition and Uses.......... 157

## CHAPTER XII.

### HOMBURG AND NAUHEIM.

Frankfort—The Stadt Darmstadt—A night with Young Germany—The City of Frankfort — The Great Fair — HOMBURG-ON-THE-HILL — Its Topography—The Cursaal and its appropriate service—Number of visitors—The Springs — Their Doses — Physiological effects — Diseases in which they are resorted to—Gouty Dyspepsia—Hypochondriasis—Scrofulous Affections—Varicose Veins, Hysteria, and Diseases peculiar to Women—Duration of the Course—The Baths—Excursion to NAUHEIM—Description of this Watering-Place—Its Springs and their Medical Use............................................................. 166

## CHAPTER XIII.

### KISSINGEN, BOCKLET, AND BRUCKENAU.

Early rising—Midnight review of a sensation novel—Departure from Frankfort—The vineyards of the Maine—Bavarian railways—A Turkish missionary—WURZBURG—Schweinfurth—The Hotel Rabe—Mountain drive—Bavarian politeness — KISSINGEN — Description of the town —The mineral springs—The Spoolenbrunnen—Remarkable phenomenon—Medicinal qualities of the waters—Visit to BOCKLET and BRUCKENAU—Account of these watering places—Their sources and their properties ... 183

## CHAPTER XIV.

### THE JOURNEY TO CARLSBAD.

Kissingen to Bamberg—A Bavarian buffet—An apparition—Hof—Journey to Zwickau—Schwartzenberg—Disinterested benevolence—Drive to Carlsbad—Austrian frontier—Attempt at Robbery—Arrival in CARLSBAD .................................................................... 200

## CHAPTER XV.

### CARLSBAD.

Its situation—The warm river—Description of the town—Amusements—Hotels—Climate—Analysis of the "Kurliste"—The springs—Account of the Brunnens—Action of the waters—Author's experience—A remarkable case—The Sprudel *versus* Mr. Banting in the treatment of obesity—Chronic hepatic disease—Irregular gout and rheumatism—Dyspepsia and hypochondriasis—Cases for which Carlsbad is unsuited .................................................................................................. 206

## CHAPTER XVI.

### MARIENBAD AND FRANZENSBAD.

Journey from Carlsbad to Marienbad—Bohemian Scenery—Character of the Peasantry—MARIENBAD—*Coup d'œil* of the Town—Number of Visitors—Hotels, Doctors, and other *Agrémens*—Analysis of the various Mineral Sources—The Baths—FRANZENSBAD—Its Topography—The Springs and Mud Baths—Their Composition and Medicinal Effects .................................................................................................. 220

## CHAPTER XVII.

### TEPLITZ AND THE MINOR BOHEMIAN SPAS.

The town of TEPLITZ and its environs—How to arrive there—The mineral waters—Their analysis and action—SCHÖNAU and its thermal sources—BILIN, "the German Vichy"—Diseases in which resorted to—The plain of the bitter waters—PÜLLNA—The village—Composition of the water SEDLITZ—The wells—Analysis of, compared to our *Seidlitz* powders—SAIDSCHÜTZ—Observations on the therapeutic influence of the Bitter waters .................................................................................................. 235

## CHAPTER XVIII.

### CANNSTADT AND WILDBAD.

CANNSTADT: its Position, Population, and Environs—A German Festival and our Visit—The Mineral Springs—Their Composition and Effects—Class of Patients by which they are used—Journey to Wildbad—The Black Forest and Valley of the Enz—A troublesome Companion—Arrival in WILDBAD—Hotel Klump—Dr. Haussmann—Description of this Spa—Detailed Account of the Bathing System—The Hospital—Curious Privilege—Geological Formation—Recent Analysis of the Water—Its Mode of Action explained—Newly-discovered Sources—Physiological Effects of the Baths—Spa Fever—Diseases in which Wildbad is resorted to .................................................................................................. 245

x                                  CONTENTS.

## CHAPTER XIX.

### BADEN-BADEN.

Its Topography—Its Advantages and Disadvantages—Descriptions of the old and new Towns—Hotels and Lodgings—Number of Visitors—The "Conversation-Haus"—A strange Scene—Society in Baden—The water as "improved"—The Mineral Springs—Composition of the Ursprung—Medicinal Action of the Water—Observations on its Employment in various diseases ............................................................................ 258

## CHAPTER XX.

### THE SWISS BADEN.

A painful operation—Situation of BADEN-ON-THE-LIMMAT—The town—Mineral sources—Their neglected condition—Analysis of the water—Ancient *renommè* of the Spa—Spa life in Baden in the fifteenth and seventeenth centuries, as described by eye-witnesses—Cases in which this spring is now employed ................................................................ 266

## CHAPTER XXI.

### SCHINZNACH, WILDEGG, AND PFEFFERS.

Route from London to SCHINZNACH—History and Description of this Spa—Composition, Properties, and Medicinal Uses of the Sulphurous Water—The Iodated Spring of WILDEGG—Medico-chemical Observations on it—Journey from Zurich—The Lakes of Zurich and Wallenstadt—Ragatz and its Thermal Establishment, Author's experience of—Walk from Ragatz to PFEFFERS—Description of this Watering Place—Its situation and romantic Scenery—Remarkable position of the Hot Springs—History of Pfeffers—Analysis of the Springs—Their Character and Uses—Accommodation and Inducements for Invalid Visitors. 273

## CHAPTER XXII.

### AIX-LES-BAINS AND VICHY.

Situation and Description of Aix—Thermal Establishment—Remains of Roman Baths — Hospital — Thermal Caves — History of the Spa—Number of Visitors—The Mineral Waters—Diseases in which they are prescribed—The Springs of MARLIOZ—Journey from Aix by the Rhone—VICHY—Account of the Town and its Environs—Accommodation for Invalids—The Bath-houses, Ancient and Modern—Geological Formation of the Country—General Observations on the Vichy Waters—Analysis of the Springs—Remarks on their use in various diseases—Manner of employing them, and regimen to be observed during the course—CUSSET, and its Alkaline Spa—Th Volcanic region of Auvergne—Mineral Springs of Central France—MONT DORE, ST. NECTAIRE, and ROYAT ......................................................... 290

## CHAPTER XXIII.

### A VISIT TO THE SPAS OF THE PYRENEES.

Reminiscences of a walking tour in the High Pyrenees—The Pleasures of Hardship—General observations on the hot Springs of these Mountains—From Tarbes to Bagnères—The Bearnèse peasantry—BAGNÈRES-DE-BIGORRE: its waters, their History, and Action—Our Walk resumed—The Road to Lourdes—Valley of Argelez—The Poetic *versus* the Prosaic—Arrival in CAUTERETS—Accommodation and Resources—The Mineral Springs examined—Diseases in which they are prescribed. ... 313

## CHAPTER XXIV.

### THE FRENCH WATERING PLACES CONTINUED.

Journey to Lux—BARÈGES—The springs—ST. SAUVEUR—Account of this Spa—BAGNÈRES-DE-LUCHON, and its mineral sources—AMÉLIE-LES-BAINS—A long walk—Midnight in the High Pyrenees—Arrens—A passage under difficulties—Arrival in EAUX-BONNES—The waters and their medicinal uses—EAUX CHAUDES—The mineral sources—Termination of our walking tour—PAU—The Chalybeate water of the Parc—DAX—"La Fontaine Chaude"—Its history and present employment—Return to Paris—The springs of PASSY, AUTEUIL, and ENGHIEN-LES-BAINS .................................................................. 328

## CHAPTER XXV.

### THE SPAS OF ITALY.

Preliminary remarks on the Italian mineral waters—The springs of Lombardy and Tuscany—ACQUI, ABANO, PISA and LUCCA—General and medical account of these Spas—MONTE CATINO—The watering places of the Roman states—CIVITA-VECCHIA, VITERBO, and PORRETTA described—Recollections of a summer in Naples—Escape to CASTELLAMARE—Advantages of this climate—The mineral springs, their analysis and therapeutic use—The island of ISCHIA—Its thermal sources and "Stufe"—Accommodation and resources for invalid visitors—The medicinal properties and mode of using these baths and waters—Conclusion ........................................................................ 354

# THE SPAS.

## PREFACE.

The favourable reception accorded to a former work by the present author, and the very indulgent notice taken of it by professional, as well as by general periodicals, emboldens him again to address the public on a kindred subject. The writer cannot but be aware that for the success of that work he was indebted, next to the kindness of his reviewers, to the circumstances which had afforded him an opportunity of studying, practically, a department of medical science which it requires a long and varied experience of many climates to treat of with advantage. The only merit which he sought to claim was that of conveying the results of his experience in as simple and as readable a form as possible; and he would rest well satisfied could he achieve a similar end in the present volume.

It would be difficult to point out a class of remedies which have so wide a range of action on disease, and which are resorted to in so great a variety of cases, as that which forms the subject of this treatise. Diseases occasioned by an inactive life, and those resulting from long continued over exertion; maladies caused by plethora, and those induced by anæmia, the impoverished blood produced by imperfect digestive or assimilative powers, and the vitiated blood laden with excess of nutriment; in a word, morbid conditions the most opposite, occurring in persons of different temperaments, of every age and of both sexes, are all under the curative influence of mineral waters.

The list of ailments benefitted by the Spas would be as long, and, in this place, as uninteresting as the catalogue of ships in the Iliad.

Most of those whom we meet at the watering places are patients suffering from some form of indigestion, or from diseases occasioned by the gouty and rheumatic diathesis. These, with cases of nervous, cutaneous, and scrofulous affections, constitute the great bulk of the Spa-drinkers. But besides persons labouring under actual disease, a large proportion of the patients at the watering places suffer from no tangible malady, although obviously " out of health."

Among the throng of pilgrims to these fountains of Hygeia, come youth and age—the fagged beauty, after a season, in search of that bright complexion and those roseate hues which cosmetics cannot counterfeit; legislators and professional men; the votaries of science and of fashion; the man of business and his more laborious rival in the race of sanitary destruction, the lounger about town,—all Hadgis to the shrine of health.

In such cases the constitution is impaired by a long perseverance in a faulty mode of life; and hence it cannot be expected that physic can of itself effect the cure. Nay, in some of these cases, medicine does more harm than good, and probably the maladies of these patients are oftentimes aggravated by the repeated dosing for which many Englishmen have so unaccountable a predilection. Such patients are the greatest difficulty of the practitioner. If what they consider sufficient physic is not prescribed at every visit, they think themselves neglected; and either resort to the advertising quack, whose pills and drops promise a certain cure in every case, or else have recourse to the no less dangerous pseudo-scientific charlatan, the homœopathist, or hydropathist. And when we compare the cautious diagnosis, guarded prognosis, and oftentimes tentative practice of the educated physician, who anxiously watches nature, and tries to lead, not to force, her back to health, with the bold practice of the quack, who, with unscrupulous audacity, vouches for the unfailing success of his nostrum, we can hardly be surprised at the progress of charlatanism of every kind.

To return to the mineral waters,—these agents, quite irrespective of their unquestionable active properties, sometimes cure disease by giving nature a respite from medicines; for as an old English poetic divine well expresses it :—

> "To make a trade of trying
> Drugs and doses, always prunning,
> Is to die for fear of dying;
> He's untuned that's always tuning.
> He that often loves to lack
> Dear bought drugs, hath found a knack
> To foil the man and feed the quack."

Another advantage of prescribing mineral waters is, that an opportunity is thus given for taking the invalid away from those habits of life which are often closely connected with his ill-health, and which, unless they are interrupted and abandoned for a time, will prevent the patient's cure. Besides this, if a distant watering place be selected, the journey, the change of climate, of scene, and of living, often exercise a most potent therapeutic action of themselves. Lastly, it supplies an opportunity for mental rest, which is not idleness, but change of thought.

Thus, to send a barrister or physician in large practice, a politician or a literary man, suffering from protracted intellectual labour, and tension of mind to the country for rest is a great mistake. Such an individual would but experience Seneca's reflection that— *otium sine literis mors est, et vivi hominis supultura.* Homer has well painted the effect of idleness when he introduces his hero "eating his heart," because unable to fight. And in a passage in that most charming book—"*The Letters of Charles Lamb,*" we find the great humorist thus depicting the effects of such inaction on himself—" Yet, in the self-condemned obliviousness, in the stagnation, some molesting yearnings of life, not yet quite killed, arise, prompting me that there was a London, and that I was of that old Jerusalem. In dreams I am in Fleet Market, but I awake, and cry myself to sleep again. I die hard, a stubborn Eloisa in this detestable garden. What have I gained by health? Intolerable dullness. What by early hours and moderate meals? A total blank."*

In short, as old Burton has it:—" Mind and body must be exercised,—not one, but both, and that in a mediocrity; otherwise it will cause a great inconvenience."† And for this purpose a journey to some distant Spa is oftentimes the best remedy for many of the anomalous complaints attendant on city life. Even if the watering places had no other recommendation, this would be sufficient. " It is asked," said Sydney Smith, " if the object can be of such importance. Perhaps not, but the pursuit is. The fox when caught is worth nothing; he is followed for the pleasure of following." Substitute the word utility for pleasure in the last sentence, and we might say the same of several mineral waters.

With regard to the plan of this work, I need hardly say much. In the first place I have endeavoured to

---

* The Works of Charles Lamb; edited by Talfourd. p. 70. London, 1846.

† Anatomy of Melancholy, Part II, Sect. 2.

lay before my readers a succinct view of the nature and uses of mineral and thermal waters. I have next described the various forms of disease that may be treated by this remedy, and pointed out the mineral water adapted to each particular case.

In the second part, I have given a detailed account of the principal watering-places of Europe, from my personal observations and notes, taken at each Spa. In doing this, I have sought to combine all the medical information essential for the prescribing physician, with those local particulars necessary to the invalid traveller.

For the narrative and descriptive portion of this work, by which I have attempted to extend its interest beyond the class who would peruse a purely professional essay, I think no apology is necessary. But if it were, I might cite the example of other writers on mineral waters. Moreover, I can see no reason for the opinion of those who think that a medical writer, as such, is precluded from making those general observations which his acquaintance with human nature under various phases, his education, and those opportunities which his profession often give him of studying men and manners more intimately than other travellers, afford.

# INTRODUCTION.

## ON THE NATURE AND USES OF MINERAL AND THERMAL WATERS.

What Constitutes a Mineral Spring—Cold Mineral Waters—Their Origin—Carbonic Acid Gas in them—Its Influence in Mineral Springs—Thermal Sources—Their Mode of Formation in the Earth—Their Connection with Volcanic Action—Classification of Mineral Springs—The Muriated, Saline, Acidulous, Alkaline, Chalybeate, Iodated, and Bitter Waters of Europe—Their Nature and Effects—Various Theories—Author's Opinions—Influence of Change of Climate, Scene and Life.

Mineral springs are those that contain saline ingredients in such quantities, or in such combination, as to possess medicinal properties. The mere amount of salts dissolved in any spring does not, of itself, make it "mineral" or not; and several important Spas contain absolutely less mineral matter than our ordinary drinking water. The various sources that supply London are all rich in saline constituents. Thus the New River water contains $2\frac{1}{2}$ grs. of solid matter to the pint; the East London Company 3 grs.; and the Hampstead Company's water $4\frac{1}{2}$ grs. Yet these waters cause no apparent effect, whilst the springs of Wildbad, with $3\frac{1}{2}$ grs. of salt to the pint; Pfeffer's and Gastein, 2 grs.; and Chaudfontaine, with $2\frac{1}{4}$ grs.;

are all capable of producing the most striking remedial action.

Mineral springs are either cold or thermal. The cold mineral springs contain a larger amount of gaseous and saline ingredients than the thermal. They have generally a comparatively superficial origin in subterranean streams, formed by the atmospheric water absorbed by elevated mountain districts; and are forced up to the surface by hydrostatic pressure; dissolving and becoming charged with the soluble salts contained in the various strata they percolate in the transit. Many Spas, however, contain gaseous and other constituents that do not exist in any geological formation; and which, therefore, must originate in chemical decomposition. The chief of these substances are sulphurated hydrogen and carbonic acid gases. The first is produced by the hydrogen of the water uniting with some sulphuret, generally that of iron, in the passage through pyrites; and the latter is formed by the action of water on limestone or carbonate of lime, a substance containing two-fifths of its weight of carbonic acid. This carbonic acid is found in mineral waters in two forms, either in combination with certain bases, such as iron, lime, or soda, with which it forms soluble salts; or it exists free in the form of gas, which escapes when the water is exposed to the atmosphere.

All water contains more or less of this gas, the amount being modified to some extent by the temperature and specific gravity of the fluid, and being greatest when both are low; for heat drives off the gas, and it escapes most readily from a highly saline water.

Carbonic acid is one of the most important constituents of the Spas. It not only produces physiological effects itself, but also renders other substances, of themselves insoluble, capable of being dissolved, and moreover, makes springs that would otherwise be mawkish and unpalatable, palatable and agreeable. Chalybeate Spas, for example, are resorted to, or neglected, not so much on account of the amount of iron they contain, as on the volume of carbonic acid gas by which it is rendered digestible and active. Thus Dr. Althaus tells us that " the strongest chalybeate known in the whole world is the Aqua Ferrata di Rio, in the island of Elba; but as it does not contain any carbonic acid, it is entirely useless for medical purposes."*

Thermal, or warm medicinal springs, are those whose temperature varies from 80 deg., as at Schlangenbad, to 165 deg., as at Carlsbad.† Many theories have been advanced to account for the phenomena of water at an elevated temperature, springing up in the immediate vicinity of cold springs, and, perhaps the most satisfactory of these is that which supposes their connexion with volcanic action. If this conjecture be correct, the

---

* On Carbonic Acid in Mineral Waters. Dublin Quarterly Journal of Medical Science, No. LXV., p. 75.

† I may here observe that in all cases in which allusion is made to degrees of temperature in the following pages the scale of Fahrenheit is referred to.

thermal waters probably issue from caverns in proximity to a deep-seated, internal source of heat, occupying the centre of our planet. Regarding these cavities as air proof boilers, so to speak, we know that under such circumstances water may be charged with caloric until, as Mr. Scrope asserts, " it becomes absolutely red hot ; " but the moment " that an opening is made in the enclosing vessel, reducing the pressure to that of the atmosphere, it flashes into steam with explosive violence." * Nothing can more closely resemble the escape of steam from a kettle of boiling water than the *per saltum* jets of the Geysers of Iceland, or the Sprudel of Carlsbad, for example ; but in these cases the kettle, so to speak, is so remote that it is condensed vapour, or boiling water, and not steam, that escapes. In the majority of cases the boiling steam, on its way to the surface, probably passes through ordinary springs, by which it is diluted and modified.

Wherever hot springs abound, we find traces of volcanic action. In some places thermal springs issue in the immediate vicinity of active volcanoes, as in the case, for instance, with the springs of Castelamare, St. Lucia, Ischia, the Geysers, &c. In other cases, where the warm springs are, perhaps, thousands of miles distant from any volcano, we still find proof that volcanic agency once played an important part in the configuration of the neighbouring country. Such are the indications given by the basaltic and granite formations

* Mr Scrope, M.P., "On Volcanoes," p. 39. London, 1862.

from which the brunnens of the Taunus mountains, the baths of Pfeffers, and the warm springs of the high Pyrenees issue. A still more striking proof of the relation between volcanic action and thermal springs is furnished by the fact that direct communication seems to have been clearly established between earthquakes occurring in one country, and the hot springs of distant lands. Thus, during the greatest earthquake of modern ages, that of Lisbon in 1755, on the 1st of November, " the water of the chief spring at Teplitz, in Bohemia, between eleven and twelve midday," a contemporary writer tells us, " first turned thick, and then became as red as blood; but after a short interval returned to its former colour, and flowed with much greater rapidity than before." Two days later, on the 11th, it was again recorded: " the waters of Teplitz turned red as blood." * During the same time the mountain above the thermal fountain of Naters in Savoy, opened and discharged hot water; and the springs of Moutiers and Leuk ceased to flow while the earthquake lasted, after which they again re-appeared, but considerably altered in character.

Mineral waters are classified into groups which bear some general resemblance, either in chemical composition, or in therapeutical action. Numerous as have been the divisions founded on either of these principles, none seem to be free from error; for if we adopt the chemical classification, we will find that a considerable

* Gentleman's Magazine. Supplement for 1755, p. 588.

number of the Spas are so complex in their character, that it would be difficult to say which ingredient so proponderates as to determine their nomenclature on chemical principles. And, moreover, we find waters containing only a trace of any chemical ingredients, producing powerful effects, while others, holding a much larger proportion of the same constituents, will sometimes occasion very different results, or even none. If, on the other hand, we base our plan of classification on the therapeutic properties of the Spas, we shall be equally liable to error, since the same waters occasion widely opposite effects, according to the doses they may be taken in, the condition of the patient, and a great variety of other circumstances. These points will all be considered in the chapters treating of the various Spas and their special indications.

Therefore, putting aside a strictly chemical classification of mineral waters as chimerical; and for the reasons I have already given, passing over, for the present, the division sometimes adopted of the Spas from their medicinal effects,—I shall consider the various springs described in the succeeding chapters as divisible into chalybeate, sulphurous, salines (simple, alkaline, muriated, and acidulous), bitter waters, iodated, earthy, and chemically indifferent mineral springs.

*Chalybeate waters* are those whose principal ingredient is the carbonate of the protoxide of iron dissolved in water containing more or less carbonic acid gas. Chalybeate springs may be divided into simple and saline. Most of them hold a certain amount of manganese in solution, and in a large proportion of these Spas, the iron is combined with saline ingredients. The action of waters containing little else than carbonic acid and iron is stimulant and tonic, exciting the nervous, circulating, and digestive functions, and at the same time improving the quality of the circulating fluid, by increasing its red corpuscles and fibrine. These Spas are therefore indicated in the treatment of purely anæmic cases, and diseases depending on poverty of the blood, cases of convalesence from fever and other acute diseases, disorders of the nervous system, especially hysteria, chlorosis, and chronic complaints, depending on derangement of the uterine functions, neuralgia and chorea.

The principal simple chalybeates are—Pyrmont, in the principality of Waldeck, formerly the most celebrated Spa in Germany, which contains nearly half-a-grain of iron in the pint of water; Driburg, in Westphalia; Spä, in Belgium; Schwalbach, in Nassau; and Brückenau, in Bavaria.

The saline chalybeate Spas which principally contain the soluble salts of soda, together with the iron, are chiefly employed in similar cases to the saline springs when marked with debility, and in anæmia, complicated with abdominal disease, whether resulting from long residence in tropical climates, or from dietetic errors. In the following pages the reader will find an account of the chief waters of this class, and their particular effects. Amongst saline chalybeate springs are included the Stahlbrunnen of Homburg, Bocklet in Bavaria, Franzensbad in Bohemia, Cronthal, Rippoldsau, Tunbridge Wells, and Cheltenham.

*Sulphurous springs* are those whose chief constituents are sulphuretted hydrogen gas and metallic sulphurets, generally of sodium or potassium. In appearance, they are usually clear or milky, and have more or less of the "rotten egg" flavour peculiar to sulphuretted hydrogen. Sulphuretted springs differ materially in their action, according to their temperature, being most powerful when this is highest.

Warm sulphurous waters, are, without exception, stimulant in their effects, producing excitement of the nervous as well as of the vascular system. The principal sources of this class are Aix-la-Chapelle, Aix-les-Bains, Baden in Switzerland, Baden near Vienna, Burtscheid, Pystjan, Schinznach, Teplitz, and Warmbrunn. The Pyrenean mineral springs nearly all belong to the same division of thermal waters, and amongst the latter the most important sources are Amélie-les-Bains, Ax, Bagnères de Luchon, Barèges, Cauterets, Eaux-Bonnes, Eaux-Chaudes, and St. Sauveur. The cold sulphurous springs, resorted to by British invalids, are neither so numerous nor so active in a therapeutic point of view as those just mentioned. Among these watering-places are included Eilsen, Langenbrücken, Nenndorf, Weilbach, and Wippfeld, in Germany; Lucan and Swadlinbar, in Ireland; Enghien-les-Bains, in France; Bex, in Switzerland; Moffat in Scotland; and Harrowgate, in England.

The physiological action of sulphurous waters is, as I have already said, stimulant, while their therapeutic effects may be best expressed by the term alterative. The most recent physiological experiments with these waters lead to the conclusion that they increase the elimination of carbonic acid from the lungs and skin, and of uric acid and urea from the kidneys. The action of these Spas is more intimately dependent on their temperature than on their chemical composition, being more stimulant the hotter they are. The warmer springs are chiefly used in diseases dependent on the gouty and rheumatic diathesis, in cutaneous eruptions, especially acne, pityriasis, sycosis, prurigo, psoriasis, syphilitic eruptions, neuralgia, glandular and arthritic swellings, abdominal plethora, and old gunshot wounds. The cold sulphurous Spas, though less powerful, are used in analogous cases to the former. Such mineral springs, however, whether hot or cold, should never be used when there is a tendency to inflammation or even congestion of any important organ, especially of the brain or lungs; and they must be most sedulously avoided in all cases in which hæmorrhage, more particularly cerebral or pulmonary, is to be apprehended.

*Muriated, Saline* Springs are those whose active constituent is chloride of sodium, or common salt. Wiesbaden, Baden-Baden, Bourbonne-les-bains, Canstadt, and Soden are thermal waters of this class; and Kissengen, Homburg, and Cheltenham are cold. Such Spas, when used internally, stimulate the gastro-intestinal mucous membrane, act upon the bowels, augment the amount of urea discharged by the kidneys, increase the elimination of effete tissues, and sharpen the appetite. They are employed in the treatment of gout, rheumatism, scrofula, dyspepsia, habitual constipation, plethora, and enlargement with torpidity of the liver.

*Saline, Alkaline* Waters chiefly contain sulphate and bicarbonate of soda. Carlsbad, Marienbad, Franzensbad, and Tarasp are examples of this class of mineral springs. Their action is purgative and deobstruent, but not, as is generally supposed, diuretic; under their use the weight of the body steadily diminishes, while the elimination of urea and uric acid is retarded. These waters are largely prescribed in abdominal plethora, habitual constipation, hepatic enlargement and inaction, and gouty dyspepsia. Carlsbad seems also to merit a trial in some cases of that intractable disease—diabetes mellitus.

*Muriated, Alkaline* sources are those in which chloride of sodium, with carbonic acid gas and bicarbonate of soda are the main ingredients. Ems, Salzbrunn, and Selters, belong to this category, nearly all of which are cold waters, with the exception of Ems. They promote digestion, are gently diuretic, aperient, and anti-acid, especially neutralising uric acid when present in the circulating fluids. Their use is indicated in the gouty diathesis and dyspepsia, certain diseases peculiar to women, hysteria, renal and hepatic affections, and they are also prescribed in some cases of chronic bronchitis.

*Simple Alkaline* springs contain bi-carbonate of soda with excess of carbonic acid. This is the characteristic of Vichy, Fachingen, Geilnau, and Bilin. By the use of these Spas the blood is rendered more alkaline, the renal secretion becomes so too, the various excretory functions, especially diuresis, are increased, and the appetite is sharpened. They are chiefly used in gout and rheumatism; also, though not so extensively, in dyspepsia depending on gastric acidity, and in renal calculi and gravel connected with the uric acid diathesis.

The *Bitter-Waters* are those containing a large proportion of the sulphates of soda and magnesia. The strongest of these is Püllna in Bohemia; in the immediate vicinity of Püllna are two other powerful bitter waters, namely Saidschütz and Sedlitz, and in the Duchy of Saxe-Meiningen is Friedrichshall, the best known in England of this class of waters. These springs, which are also known in Germany as "purging, or fever waters," act generally as saline cathartics, and are also diuretic and derivative. They are principally employed in general and local plethora, obstruction and torpor of the intestinal canal, some cases of obesity, in scrofula, especially when connected with glandular enlargements, in tendency of blood to the head, and thoracic and cerebral congestions.

*Jod-und-Bromhältige Kochsalzwässer*, as the Germans call springs containing iodine and bromine, generally owe their properties to iodide of sodium and bromide of manganese in a muriated saline water. Nearly all such sources are cold. The most important waters of this class are Kreuznach on the Rhine, Adelheidsquelle in Bavaria, Hall in Austria, Elmem in Prussia, Wildegg in Switzerland, and Salzhausen in Hesse Darmstadt. Taken internally, these springs stimulate the mucous membranes, occasion ptyalism and diuresis, promote absorption and quicken the appetite. They are ordered in scrofulous diseases, especially of the glands and joints; in chronic glandular enlargements, such as bronchocele, in secondary syphilis, and some cutaneous affections.

*Earthy Springs* are impregnated with sulphate, carbonate, and chloride of lime, and carbonic acid. They are in general thermal. Wildugen, Leuk, Lippspringe, Weissenburg, Bath, Lucca, and Pisa, belong to this class of waters, the principal action of which is astringent and stimulant. Wildugen, which contains free carbonic acid, is also diuretic, and is useful in gravel and diseases of the bladder. Leuk is chiefly resorted to by patients suffering from skin diseases.

Bath, Pisa, and Lucca are used in gout, rheumatism, cutaneous affections, and dyspepsia

The so-called *chemically indifferent* springs are all mineral waters, an imperial pint of which contains less than four grains of saline matter. Nearly all these fountains are thermal, and to a great extent owe their virtues to this fact. The principal Spas belonging to this category are Schlangenbad, in Nassau; Teplitz, in Bohemia; Gastein, Neuhaus, and Tüffer, in Austria; Wildbad, in Wurtemburg; Pfeffers, in Switzerland; and Chaudfontaine, in Belgium. These waters, though so weak in chemical ingredients, are yet powerful remedial agents, and are principally used for bathing purposes. Their effects depend on their temperature, the cooler ones being calming or sedative, allaying nervous irritation, and diminishing vascular excitement. Some of them, such as Wildbad and Schlangenbad, afford peculiarly agreeable sensations to the bather, and render the skin soft and white. These are employed in rheumatism, rheumatic arthritis, impeded menstruation, and neuralgia. The hotter springs, such as Teplitz and Gastein, are used in more aggravated cases of chronic rheumatism and consequent loss of power. They are more exciting and stimulant, and if prolonged too much, produce headache and febrile disturbance.

We may now proceed to consider the practical application of these various mineral waters to the cure or alleviation of disease, which has much less connexion than is generally supposed with their chemical composition, although able chemists, such as Dr. Aldridge, tell us that it is only necessary to ascertain what are the saline ingredients in any mineral water, " and dissolve the same constituents in the proper proportions, and you regenerate the Spa to all intents and purposes."* I presume, however, that the "intent and purpose" of the invalid who resorts to any Spa is the cure of some ailment; and most assuredly this will not be accomplished by the "simple method" of the chemist. The baths of Wildbad, for instance, contain only two grains of common salt, with half a grain each of carbonate and sulphate of soda to the pint of water, at the temperature of 98 deg.; and yet, wonderful cures of chronic arthritic diseases are effected by a course of them. But very few practical physicians could be persuaded that an artificial solution of the same salts would produce the same effects. If it had the same influence, would it not be as absurd as cruel, to send a patient crippled by chronic rheumatism to a remote village in the Black Forest, when a handful of common salt, with a teaspoonful of washing soda and Epsom salts, would convert a tub of luke-warm water in his own bedroom into a *verjungen*, or "youth restoring fountain?"

A great many theories have been advanced to account for the therapeutic action of mineral springs,

* The German Spas and Vichy. By T. Aldridge, M.D. p. 185. Dublin, 1856.

especially by German writers, who seem to delight in explaining the *ignotum per ignotius*, and almost every succeeding writer puts forward a conjecture, if possible, more misty and unintelligible than his predecessor. The following specimens from German, French, and English Spa authorities will suffice to illustrate the peculiar style of these opinions. Dr. Gugert, a resident physician at Baden-Baden, thus writes :—" There is, probably," he says, " an electric galvanic fluidness which imparts the same in a bulk, by forming the springs in the interior of the earth, through an electric volcanic process ; and which does especially contribute to restore the disturbed harmony in the sick human organisation."* Dr. Peez, of Wiesbaden, attributes the cures performed at the watering places to " a peculiar vital principle in mineral waters."† Dr. Granville ascribes the same result to a supposed peculiar " thermal or telluric heat," existing in mineral waters, and which he believed to be " specific in its action, and therefore dissimilar from ordinary heat."‡ Very recently, in a communication to the French Academy of Sciences, M. Scoutetten asserts that :— " All Spa waters excite an electric current when in contact with the animal tissues, the current varying in intensity according to the nature of the water. A feeble current may be produced

---

\* Dr. Gugert—Treatise on Baden Mineral Waters, in Beauties of Baden-Baden. p. 45. London, 1858.
† Dr. Peez—The Mineral waters of Wiesbaden. p. 103.
‡ Dr. Granville—The Spas of Germany. 2nd Edition. p. 23.

even by river-water; but mineral waters proper give rise to currents, some of which are so powerful as to deflect the needle of the galvanometer from 80 to 90 degrees. These currents are said to traverse the body, and produce a medicinal effect; but we have no information as to the special mode in which the effect is produced, nor in what it differs from that of electric currents generated by other means."

I have little faith myself in any of these theories. Some of them are simply absurd, while even the most rational cannot be proved, and would explain nothing if it were. The truth is, that, as is the case with most of the remedies employed in the practice of physic, our knowledge of mineral waters is merely empirical. We know, by experience, that certain waters produce certain effects, and oftentimes we must rest content with this knowledge, nor vainly trouble ourselves to search out their *modus operandi*.

My own opinion is, that in most cases mineral waters act only as diluents, and produce their good effects simply as so much pure water would do, could it be swallowed at the same temperature, under the same circumstances, and with the same anticipation of success which mineral waters are used with. When we consider the amount of water consumed daily by each patient at the Spas, we will find, that as diluents, these waters must have a considerable action on the animal economy. Ten or twelve pints of mineral water per diem is a common dose at several of the Spas, and it must be obvious that the mere passage of so large a quantity of fluid carried by the circulation through every organ and structure of the body, must break down and wash away morbid deposits and disordered secretions. Thus, for instance, when free lactic or lithic acid, or lithates, abound in the blood, giving rise to rheumatism or gout, if we dilute the vital fluid, we increase thereby the excretions from the kidneys and skin, and thus eliminate the *materies morbi* from the system, and so with many other diseases.

The effects of the journey to the Spas are very often sufficient of themselves, without any reference to the waters, to account for nearly all the benefit ascribed to the mineral springs. The majority of invalids who resort to the watering places of Germany or France, suffer from disorders connected, though perhaps obscurely, with the gouty diathesis, or with some gastric derangement. Now, as I showed in my work on " *Change of Climate*," we possess in that measure the best remedy for such ailments. For, travel as we will, whether we "rough it" on foot through the mountain bye-roads of Switzerland or Spain, or sedulously avoid fatigue in the coupé of a train, however soft may be the cushions and easy the springs of the carriage; still, more or less exercise must be taken; and it is this exercise that does good. By it the circulation is quickened, but at the same time is equalised, that is, all the vessels receive more blood, and thus the amount accumulated in any one organ is *pro-tanto* diminished.

The respiration is also hurried; and therefore, more carbon is exhaled from the system. The appetite is sharpened, and persons travelling notoriously eat far more than they do at home. The change of diet, the freer use of fruit, the light acidulous wines, the oleaginous cookery, all promote the alvine and renal excretions. The "moral" influence of the change too conduces to the physical improvement it effects. New scenes and places suggest new thoughts; the *atrabilis* of gloomy apprehension is purged away, and the patient, ceasing to think on his symptoms, they cease to exist.

Still I am very far from asserting that change of living and scene will of itself produce all the benefits often effected by a journey to, and use of an appropriate Spa. No amount of travel will, unaided by other treatment, purify the vitiated blood of a gouty patient, reduce to proper proportions a tumefied liver, unbend a contracted joint, or impart vigour to a palsied limb, all of which cures are oftentimes performed at the Spas.

Though nearly all the remedial substances contained in mineral waters are articles of the Materia Medica, and generally exist in comparatively small quantities in the Spas, yet they act more effectually in them than in any of our pharmaceutical preparations. This is probably owing to a more perfect solution, and finer division of the natural preparation, than can be effected by artificial means. Thus the active principles of the mineral waters are more readily imbibed by the absorbents into the circulation, and find admission into the most minute capillary vessels, into which our coarser medicines never enter. Another reason of this superiority is, that Spas occasionally contain substances insoluble in ordinary water. The Homburg springs, for instance, contain carbonates of lime, iron, and magnesia, and silica, which are all insoluble in common waters, but which are in this case rendered soluble by the presence of free carbonic acid gas.

The mode of using, and dose of the Spas, depends on the composition of the particular spring selected, the condition of the patient as to age, sex, temperament, his ailment, and a variety of circumstances, which render it impossible to lay down any accurate general rules on these points. Still I shall attempt what I consider essential, viz., a few general suggestions which may be found of some importance to invalids who resort to mineral and thermal springs.

In the first place, mineral waters should, if possible, be drank at their source; for the same good effects are seldom produced by the waters even at a very short distance from their spring. The patient who uses the imported mineral waters sold in this country, not only loses the chief benefit of the prescription, that is the change of scene, of climate, and of living, and all the advantages derivable therefrom, but further the fluid he drinks has no longer the same properties it possessed when bottled at the Spa, being, in most cases, greatly changed and deteriorated by keeping. And I know from my own experience that, even when no chemical change can be detected, the effect of an imported, is often very different from that of a fresh mineral water. Secondly, before the patient is sent to the Spa he should generally go through a course of laxative and alterative medicine.

The Spa physicians, as a rule, advise their patients to disregard all 'other medicines, and confide in the virtues of the mineral water of their place of abode. It would be rather difficult to say whether this recommendation has done most harm or good. In many cases it is, doubtless a wholesome counsel; for not a few English people, sick or well, have a remarkable love of physicking themselves; and it is by interrupting this for a time, and allowing nature a respite from the torture of incessant dosing that some of the Spa doctors in reality cure their patients. But, on the other hand, when the patient is really ill, probably as much harm results from the neglect of necessary remedies. Ptisans and placebos are all very well in their way, as long as it is only necessary to amuse the patient while nature cures the disease, but a serious malady must be met by serious and active treatment. The use of mineral water should not, I think, supersede all other remedial measures, for, as a very observant English physician of the last century well remarked, "It is but prudent to bring all the forces one can raise against so formidable and so potent an enemy as a chronical distemper."*

The quantity of water to be consumed is a matter that should be prescribed to the patient before he is sent to the Spa, and must be determined by the circumstances of each case. For invalids are generally left to their own discretion on that point, and seldom exercise any. Not a few valetudinarians think that the benefit to be derived is in exact proportion to the amount of Spa water they can swallow, and accordingly load the stomach with so much fluid, that they weaken and overpower, instead of repairing the digestive functions. Generally speaking, two glasses of water before breakfast, one before dinner, and a couple of tumblers-full in the evening are quite sufficient.

The mode of life at the Spas is usually totally different from the ordinary habits of the patient, who is forced into earlier and more regular hours, and into taking more exercise than he had been previously accustomed to, and this change is probably one of the most important advantages of the foreign watering places. As, however, in the next chapter I shall enter into this subject at some length, I need here only observe that the water-drinker should live abstemiously, and take as much exercise as his strength will allow of without over fatigue.

* Cheyne. "An Essay on the Gout and the Nature and Quality of the Bath Waters," p. 59. London, 1721.

It should be clearly understood that there are many morbid conditions and diseases in which mineral waters cannot possibly do good, and may probably do positive injury. This seems a very trite observation; but it is so often neglected, that I deem it essential to impress on my readers that this class of remedy is not suited to the treatment of organic disease. When structural change has taken place in any important organ, mineral waters are worse than useless. Yet consumptive patients are constantly sent to certain Spas, and we are told by the local authorities that cures are thus effected. Having devoted a great deal of time and observation to the study of Phthisis in many climates, and visited almost all the Spas which are alleged to be beneficial in that disease, I may state my belief, that no case of consumption was ever cured by any mineral water, though very possibly the general health of phthisical invalids may occasionally be considerably improved by a course of light chalybeate or alterative waters.

Besides their internal use, mineral and thermal springs are also employed externally in the form of baths. The action of mineral baths depends on their chemical composition and temperature, and as this varies so widely that no two of these natural thermæ agree exactly in either respect, it is obviously impossible, in these prefatory remarks, to do more than offer a few general considerations on such phenomena and effects as are common to most thermal baths, leaving the application of these observations to the chapters that treat of each of the Spas.

The usual division of natural warm baths is into three classes, according to their temperature; but as this arrangement seems to me wholly arbitrary and unnecessary, I prefer considering the thermal waters as forming two groups, the first of which have a temperature ranging from 85 deg. to 97 deg., or lower than that of the blood; and the second comprising all the baths whose heat exceeds this. The effects produced by baths of the former kind are of a sedative character; after a few minutes' immersion the nervous system is soothed, pain and irritation are allayed, a feeling of physical comfort and tranquillity is induced; the functions of the skin, however, become more active, the various secretions are increased, and changes are produced in the blood, by the transudation and absorbtion that take place in the bath, which are greatly influenced by the chemical composition and density of the water. The principal maladies treated by baths of this class are—chronic nervous and spasmodic affections, such as neuralgia and sciatica, chronic subacute disorders of the abdominal and pelvic viscera, especially of the liver and gastro-intestinal mucous membrane, leading to dyspepsia and hypochondriasis. This remedy is also applicable to cases of obstructed discharges, chronic rheumatism, and to certain, though not very many, cases of gout, and, finally, these baths are used with remarkable success in obstinate chronic skin diseases.

Almost all the chronic ailments that drive invalids abroad to the Spas, are attended with more or less pain, irritation, or discomfort; and if, by bathing in, and drinking a thermal water, and at the same time observing a suitable mode of living, this uneasiness can be allayed, or soothed; not only is the present condition of the patient thereby improved, but oftentimes a great step is made towards a permanent cure.

In many chronic diseases, especially those occurring in individuals beyond the middle age, and of inactive or sedentary habits, there is a disposition to passive congestion in some of the internal organs, accompanied with torpidity and languor of the cutaneous circulation. Such cases are generally marked by a slow pulse, deficient secretions, cold extremities, and a pale or sallow skin. Now, if a person in this state be placed in a warm bath, rendered more stimulating by its saline contents, the blood is attracted to the cutaneous vessels, relieving the internal organs of the blood thus drawn to the surface, transpiration takes place, the other secreting glands—the liver, kidneys, &c., are stimulated to increased action, and thus what the older physicians called " peccant humours "—a term I would retain, for it expresses clearly enough an idea that modern writers endeavour to convey by more euphonious circumlocutions—are expelled; the blood is thus brought to a healthy condition, and the balance of the circulation is restored. Nor is this benefit merely temporary; for a more healthy tone is gradually imparted to the blood vessels, and the vitiated secretions that are got rid of are repaired by the healthy nutrition which results.

The second class of thermal baths, comprehends all those whose temperature is hotter than that of the blood, and ranges from 98° to 112°. Such baths are strongly stimulant, their exciting power being in proportion to the elevation of their temperature. The blood is determined to the surface by them, the cutaneous capillary vessels become congested, and the skin assumes a red arterial hue, the respiration is hurried and embarrassed, the pulse is accelerated and becomes full and throbbing, a tendency to fainting is experienced, and a sense of oppression and discomfort is felt until a profuse sweat relieves the circulation. When this perspiration has broken out, all the uneasy sensations gradually disappear, and are succeeded by a sense of languor and exhaustion, with a tendency to sleep.

Widely spread, as they are throughout Europe, the therapeutic use of these hot mineral baths is limited to a comparatively small class of patients. They are principally employed in certain obstinate skin diseases, in chronic rheumatism, and rheumatic arthritis; also in cases in which it is doubtful whether certain obscure symptoms result from a specific blood disease, or are attributable to the action of mercury; in which case some hot mineral baths—as I purpose to prove in the chapters on Aix-la-Chapelle and Eaux-Bonnes—act both as a test and as a remedy.

The length of time the patient should remain in the bath depends on the nature of the mineral water, and on that of his disease, and varies from ten minutes, which is the usual period of immersion at some of the hotter sulphurous baths, to eight or ten hours, which formerly at Pfeffers was thought an ordinary occurrence. Dr. James Johnson quotes an old writer, who asserts that "Tous les Samedis on voit accourir à Pfeffers une multitude des gens des compagne voisines, et ils restent dans le bains jusqu'au Lundi matin pour provoquer la sueur."

The ordinary duration of a bath at Wildbad, Aix-les-Bains, Gastein, and several other places is an hour. So prolonged an immersion in a fluid at the same temperature as the blood maintains a powerful action on the skin, withdraws the blood from internal congested organs, allows their vessels to recover their healthy tone and contractile power, and must be capable of producing powerful curative results.

Care should be taken to avoid exposure to cold immediately after leaving the warm bath, and therefore I think that the very early hour at which the bath is generally taken at the spas of Germany is open to some objections. The baths there are almost always taken between half-past five and eight in the morning, and on coming out of the warm water the patient is generally enjoined, in the intervals of the goblets of mineral water which are usually prescribed with the bath, to take a smart walk before breakfast. I am convinced, however, that a great deal of harm is done by the sudden transition from a temperature as warm as the blood, to the cool morning air. And though, early rising and a walk before breakfast, do probably, fully as much good to the majority of Spa visitors as are effected by the internal and external use of the mineral waters, yet, I think that those who bathe at this time of day would be more served by remaining quietly within doors for some hours, than by exposing themselves with open pores and relaxed fibres to the sharp and often damp matutinal breeze. It might perhaps often be found better to take the warm mineral bath in the middle of the day, between three and four hours after breakfast, or in the evening, about five or six hours after the early German dinner.

Thermal mineral baths do not at all belong, as many seem to think, to that class of remedies which, if they do no great good, have, at least, the negative advantage of doing no great harm; but, unfortunately, are fully as powerful for evil as for good. There are quite as many morbid conditions capable of being aggravated by hot mineral baths, as diseases which may be cured by them. And cases frequently present themselves in which trivial complaints have been developed into grave maladies, or perfect health undermined by the improper use of these hot baths.

A consumptive patient, for instance, suffering from

some rheumatic or neuralgic complaint, should not be ordered, as an otherwise healthy individual might be, to any of the thermal mineral baths. The following case illustrative of this opinion came before me lately. A gentleman of phthisical diathesis, who had repeatedly spat blood, but who supposed himself perfectly recovered from any pulmonary disease, was attacked by very obstinate urticaria. For some time he was under medical treatment. Finding himself, however, still uncured, he at last "threw physic to the dogs," and, without consulting his physician, he repaired to the hot sulphurous baths of Aix-la-Chapelle, where, as soon as he arrived, he took his first and last bath. That same night profuse hæmoptysis set in, which recurred at intervals for some days, and did not cease until he was so weakened that it was with great difficulty he was brought back alive to this country. Such cases show the folly of those, who, because certain baths are recommended in gout, rheumatism, or any other disease from which they may suffer, think that they must be applicable to their own case, and, without any medical advice, set off thither.

Whenever a patient exhibits a tendency to determination of blood to the head, or apoplectic disposition, hot mineral baths are most dangerous, and might lead to sudden and fatal results. Very fat people should be cautious of indulging in such baths. For, in the first place, the tendency to obesity is thus increased, and secondly, in these cases there is frequently some fatty deposit in the heart which would render hot baths, whether mineralized or not, improper. In all cases, too, in which the heart, aorta, or any important vessel is diseased; and, indeed, whenever it is neccessary to guard against vascular excitement, thermal mineral waters are contra-indicated.

# CHAPTER I.

### DYSPEPSIA AND THE SPAS.

Nature and symptoms of indigestion—Moral and physical importance of a good digestion.—DIETETICS: A mixed diet best—Errors of ordinary habits of living—Abstinence—Rule for dyspeptics—Use and abuse of wine and stimulants—Treatment of indigestion—Benefits of travelling—Mineral waters employed in dyspepsia and hypochondriasis.

WITH the exception of the maladies produced in this climate by wet and cold, most of the diseases for which physicians are consulted are the result of an excessive or injudicious diet,—" plures crapula quam gladius." The object of the present chapter is to shew that many of these complaints may be better treated by mineral waters, conjoined to abstinence or some alteration of living, than by pharmaceutical preparations.

It would be impossible, and out of place if it were possible, to enter here into any detailed account of the various symptoms and varieties of dyspepsia; which, however necessary in a strictly professional essay, would be useless in a work intended as a guide for patients, as well as practitioners, in the selection of mineral waters. Under the term dyspepsia, therefore, I include all cases of chronic indisposition, whose leading symptoms are impaired, or fastidious appetite, painful, slow, or imperfect digestion, and irregular action, indicate generally a torpid state of the intestinal canal, and hypochondriasis or groundless depression of spirits.

The study of dietetics is much less cultivated than, from its extreme practical importance, it ought to be by all classes. And while persons justly attach great consequence to the medicines which are prescribed when they are ill, we find that little or no consideration is given to the digestibility and fitness of the food on which they habitually live; notwithstanding Dr. Arbuthnot's suggestion that, "what is daily taken by pounds must be more important than what is occasionally taken by grains."

A good digestion has more to do with human happiness than any other circumstance, excepting a good conscience, and even on that exercises no small influence.

"The longer I live," said Sydney Smith, "the more I am convinced that the apothecary is of more importance than Seneca; and that half the unhappiness in the world proceeds from little stoppages: from a duct choked up, from food pressing in the wrong place, from a vext duodenum, or an agitated pylorus . . . In the same manner old friendships are destroyed by toasted cheese, and hard salt meat has led to suicide. Unpleasant feelings of the body produce correspondent sensations in the mind, and a great scene of wretchedness is sketched out by a morsel of undigested and misguided food. Of such infinite consequence to happiness is it to study the body."*

* Memoir of the Rev. Sydney Smith, by his daughter, Lady Holland, vol. I., p. 126, London, 1855.

On no subject whatever have more numerous and contradictory systems been propounded than on diet, and on none would it be more utterly impossible to lay down any universally useful general rules than on this. The old proverb that, "one man's meat is another man's poison," is often literally true. Cases are recorded in which idiosyncrasies existed, rendering common articles of food productive of symptoms of poisoning. I am acquainted with one gentleman who cannot taste pork, goose, veal, or salmon without being thereby rendered ill for days; another in whom an egg produces gastralgia; and a third, who, if he takes crab or lobster, even the smallest quantity inadvertently in a sauce, pays the penalty of it by violent and continued retching within an hour. Such facts shew the necessity of not contemning our natural antipathies to certain things, which, although harmless to the majority of mankind are not so to all.

We are often, however, not better able to explain why this should be the case, than to account for those instinctive aversions experienced towards certain individuals, of whom we can only say with Martial:—

> "Non amo te Sabidi, nec possum dicere quare;
> Hoc tantum possum dicere, Non amo te."

Of all animals, man alone is truly omnivorous. And, although horses and sheep, for instance, have been trained to live on animal food, and the carnivora have been brought, by Spallanzani, to subsist on vegetable diet; yet such cases are but the exceptions to the general law, that the range of food, as well as of climate, within which the lower animals can vary with impunity, is comparatively limited. While man alone thrives equally in all climates, and on all foods—whether vegetable, animal, or mixed—although in this climate, at least, he certainly lives best on a mixed diet. But, even here, I am acquainted with persons who never eat animal food, and with others who live exclusively on it. Both being equally persuaded that their own system is the panacea for all the ills that flesh is heir to. For the great majority of people, however, an exclusively animal dietary would, in this climate, be a slow but effectual mode of poisoning.

When I assert that the majority of people eat too much and too often, I know I shall be told that in this respect mankind are now less gross than of old ; when, as in the patriarchal days, the hero's courage was measured by his appetite, as Homer describes ; or how great has been the improvement in this respect since the banquets of Vitellius and Apicius, at which, as Seneca assures us,—*vomunt ut edunt, edunt ut vomunt*. Though such revolting gluttony has long since been forgotten, yet people still eat far more than nature requires, far more than they can assimilate, and consequently comparatively very few enjoy perfect health. Meat is used at breakfast, luncheon, and dinner, by persons leading a sedentary inactive life ; and at each meal enough is often taken to sustain a labouring man. Two meals a day are sufficient for most persons in ordinary health, and no one should use animal food more than once daily, and then only in moderate quantities. The number of meals, and quantity consumed at each, however, must after all be decided by each individual's experience of what agrees with himself. For some men would gain flesh on Cornaro's allowance of eleven ounces of food per diem, while others would require as many pounds.

We might, however, all live on much less food than is commonly supposed necessary. There is a very exaggerated fear of the debilitating effects of even a short abstinence from food; for although Hippocrates says, that those who abstain from nourishment for seven days must of necessity die therefrom, we have numerous well authenticated cases in which this period has been exceeded with impunity. I was once consulted by a patient suffering from stricture of the osophagus, who for ten days before he came to me had been completely unable to swallow any solid food whatever, and with great difficulty could take even fluids. During that time, he had lived on a few tea spoonfuls of milk and water, and yet, notwithstanding this almost perfect abstinence, and although greatly reduced by his long continued preceding illness, he was still able to work a little on his farm, and walked in some miles from the country to the hospital to which he was sent. So prolonged a fast would, however, have undoubtedly proved fatal to the majority of people who differ materially in their endurance of fasting from the "never eating" tribe that Pliny makes mention of.

The only rule which I should recommend to every dyspeptic patient is, that, putting aside the bugbear of debility, he should use only the quantity of food he can digest, without pain or discomfort. And no matter how little that may be, at first he should confine himself rigidly to it. In severe cases of painful digestion, the sufferer should not taste animal food, but should confine himself to a bland farinacious diet. A dyspeptic invalid will oftentimes gain more strength on a cup of gruel or arrowroot digested without pain or inconvenience, than on a full meat diet, which only acts as a fresh source of irritation.

Although we have progressed vastly from the bibulous habits of our grandfathers, still, to the present day, too much wine is used by the upper classes in this country, and many of the ailments by which life is embittered, and premature death induced, are owing to this cause. Nothing, however, seems to me more absurd than the arguments of those, who because wine may be abused, therefore argue against its use, and would even restrict its employment by penal enactments. As well might the use of razors for shaving be prohibited, because they might be employed for cutting throats. It is perfectly true that wine and all other alcoholic liquors are stimulants,—in other words are medicines, and are therefore quite unnecessary to a man in perfect health; but comparatively very few are to be found in that happy state. If we lived in a world in which no sorrows, cares, or anxieties of the mind, re-acting on its frail tenement, the body, produced diseases, and paved the way to death, then would I join those who preached total abstinence from all intoxicating drinks. But as unfortunately we live on earth, and not in Utopia, and have to fight the battle of life often against unequal odds, and with failing strength, and sinking heart, therefore, with Solomon, would I prescribe—"Wine to him that is ready to perish, and to him that hath grief of heart."*

Two or three glasses of wine daily, are quite sufficient for any adult, and under no circumstances whatever should wine be given to a child, except as a medicine. It would be out of place here to enter on a disquisition on the comparative merits of spirits, beer, and different descriptions of wine, though I believe a little dry sherry will usually be found the best stimulant for a dyspeptic patient.

Wine cannot, however, be considered as *necessary* for digestion, for even those who regard it in that light, do not use it with all their meals, but usually only with one repast. The only solid objection to the moderate use of wine is, that it stimulates the stomach, creates a false appetite, and prompts people to take more than nature requires.

In theory it is better to avoid taking any liquid

---

* Proverbs xxxi, 6.

with a full meal, but in practice, many people can eat nothing unless they do so. And whenever this is the case, the individual must be guided by his own sensations and appetite, and not by any rule.

To treat dyspepsia, we must bear in mind its causes; and these, besides errors of diet, are severe mental labour, conjoined to a sedentary life, nervous excitement and anxiety, the abuse of tobacco or tea, and late and irregular hours. These must all be corrected, and a journey to, and residence in, a distant watering place will often prove the most efficacious mode of effecting this. The travelling, with its concomitant change of scene and of climate, benefits the patient's general health, takes his attention from his ailments, and thus proves an antidote to the gloomy and depressant influence of dyspepsia, which, if allowed to continue, would result in hypochondriasis. The early and regular hours observed at the continental Spas, are a wholesome contrast to the late hours of fashionable life, where the example of Smyndiris, the Sybarite, who for twenty years never saw the sun rise or set, is very generally followed. Lastly, the mineral waters themselves are remedies generally far more powerful, than those prepared by the apothecary, in the treatment of indigestion and hypochondriasis.

Most of the Spas resorted to by dyspeptic invalids are saline springs, of which Soden, Homburg, Wiesbaden, and Kissingen, especially the Rakoczy source, are perhaps the most generally employed saline waters. Occasionally the more stimulating and cathartic springs, containing the sulphates of soda and magnesia, with a large proportion of carbonic acid, such as Franzensbad, Carlsbad, and Marienbad, exert a happy influence on the ganglionic nervous system, so intimately connected with digestion, and thus cure dyspepsia. Often, when indigestion is obviously dependent on nervous debility and weakness, the chalybeate saline waters are indicated, and Pyrmont, Spa, and Ems are examples of the most commonly prescribed chalybeates in dyspepsia. Other classes of mineral springs are recommended in the various modifications of gastric derangement, but so Protean are the forms of indigestion, that it would be impossible here to enlarge on this topic. Whatever Spa may be chosen, the patient should be taught to depend more on his own self-control than on the mineral water for his cure. Exercise and abstinence are the chief remedies for dyspepsia, and mineral waters and other medicines are but adjuvants, though sometimes very powerful ones.

## CHAPTER II.

ON GOUT AND ITS TREATMENT BY MINERAL WATERS.

Preliminary remarks—Gout not confined to the fashionable world—Causes of the disease—Premonitory symptoms—Description of a fit of podagra—Letter of a gouty wit—Irregular gout and the gouty diathesis—Treatment—Opinions of the ancients—Author's method—Springs resorted to—Simple saline waters, Weisbaden and Homburg—Saline Alkaline waters, Carlsbad and the Bohemian Springs—Carbonated Spas, Vichy, Fachingen, and Bilin—The English gout-curing watering places—General rules for gouty patients.

Although an elaborate disquisition on the gout would be unsuited to a work intended for both professional and general readers, as the former might find some observations imperfect or familiar, which, to the latter, would perhaps be unintelligible, unless prefaced by such information as would convert this into a popular essay on physiology, combined to a domestic practice of physic, instead of a treatise on mineral waters; still a short sketch of the leading phenomena of the disease appears to me necessary for understanding the action of the Spas in its cure.

Formerly gout was deemed a sign of wealth, and even of a vigorous intellect. Thus Sydenham remarks that "gout destroys more rich than poor, and more wise men than fools."

Whatever may have been the case then, at present, however, gout attacks rich and poor with impartiality, and were it confined to the wise alone, the fees of gouty patients would, I fear, be "few and far between." The fact is, that whenever people of any class indulge largely in animal food and fermented liquors, gout will prevail; though, of course, the injury caused by the excessive use of azotized articles of diet will be less in proportion to the amount of exercise or labour performed. Even hard labour, however, will not always counteract the effects of too free a mode of living. Dr. Budd tells us of a class of labourers employed in raising ballast from the Thames, who are obliged to work at irregular hours of the day and night, exposed to all the vicissitudes of the weather, and undergoing great exertion, to sustain themselves under these hardships, "each man," he says, "drinks from two to four gallons of porter daily, and generally a considerable amount of spirits besides. Gout is remarkably frequent among them." But though the gout may be produced or engendered in almost any individual by certain habits of living, far more frequently it is the result of hereditary transmission. Not that the gout itself is thus bequeathed from father to son, but a tendency to gout, or the gouty diathesis descends for generations; and the inheritor of these tendencies, unless he use precautions very few submit to, when he attains middle age will experience the effects of his own and his ancestors indulgence at table, and want of exercise, in a fit of the gout.

Regular gout, unlike the irregular form, which for the most part attacks elderly persons of enfeebled and broken-down constitutions, occurs generally in middle-aged plethoric individuals.

An attack of regular gout is generally preceded by certain premonitory symptoms, of which the most constant are, gastric derangement, flatulency, and heart-burn, diminished excretions, languor, groundless depression of spirits, and great irritability of temper. When such symptoms occur in a man of full habit of body, beyond the middle age, and especially if he be of gouty parentage, we may safely conclude, unless there is some other obvious cause for them, that gout will declare itself in a few days. If no attention be paid to the warning thus given, and nothing be done to stave off the approaching seizure, after three or four days' slight indisposition the patient may go to bed in his usual condition, and awake from a sound sleep, to suffer the torture of acute gout. These premonitory symptoms are caused by the presence in the blood of an abnormal constituent, uric acid, which, as Dr. Garrod has shown,* combines with the soda of that fluid, and is deposited in the part attacked in the form of urate of soda.

The deposit of the gouty matter is attended by the

* "On the Nature and Treatment of Gout and Rheumatic Gout," p. 340. London. 1859.

following symptoms. About the middle of the night the sufferer is awakened by a violent pain in the foot, generally in the ball of the great toe. This pain at first is, as Sydenham says, "like the pain of a dislocation, and yet the parts feel as if cold water were poured over them;" shivering and feverishness succeed, and the pain increases and becomes excruciating. The suffering occasioned by acute gout has been variously designated as a "tearing," "gnawing," or "stretching pain," but the best description given of it that I know of, is that of the patient, who compared the sensation of gout to a rat with teeth of red hot iron gnawing the tendons of his foot.

After some hours of restlessness and suffering, the very intensity of which brings its own relief; worn out by pain, the patient sinks into a deep sleep, which generally lasts till late the next morning, and on being aroused finds his foot sore and tender, somewhat swollen, and very red. For some nights the pain and febrile disturbance recur, though with less severity, the cuticle peels off, the urine becomes more copious, depositing lithic salts, and the attack gradually wears away. The period which now elapses before the next seizure is termed the "interval," and after a first fit of gout this will, probably, last a year, but becomes shorter after each recurrence, till at last the invalid is hardly ever free from the disease.

As I observed in another work, a patient's own account of his feelings often convey a more accurate notion of his malady than could be given by whole pages of medical description. And, therefore, I lay before my reader the following vivid and amusing account of his own sufferings, written by a celebrated gouty wit, of the Walpolian age, Robert Jephson, the author of " Roman Portraits," and " The Confessions of Couteau," in a manuscript letter to the Countess of Mountjoy, which lately fell into my hands :—

" Black Rock, 13th July, 1795.

" My dear Lady Mountjoy,

" You may guess how ill it is with me when I find myself again obliged to deny myself the pleasure of your company next Thursday. I have had a severe fit of the gout in both my feet—it is a little gone out of my right foot, but I believe it is drawing teeth out of my left at this moment. In short, I have the pleasure of feeling as if my foot was well squeezed in an iron vice. God knows when I shall be better. I am carried about like a child of six weeks old. My greatest exploit is walking up or down stairs on my back, with the help of my arms and three servants, and this is really putting my best leg foremost. My dear friend Robert! I always thought he had wonderful patience, but now I am convinced Job was but a type of him. Nothing exasperates me so much as being wished joy of the gout—a common mode of salutation. D—— their impudence! they might as well wish me joy of the plague or a spotted fever. I threw a crutch at one fellow who wished me joy, and intend to knock him down if he comes again within my reach with the same words in his mouth. To add to my comfort my best servant has also got the gout, and poor Mrs. Jephson is worse than I am. I am sometimes a little addicted to crying, sometimes whistling, sometimes singing out of tune, greatly given to cursing, and swearing, but always, with the most sincere affection, your ladyship's faithful and most miserable humble servant,

"ROBERT JEPHSON."

Besides what is known as the regular gout, the disease in question plays a protean, and not sufficiently recognized part in the chronic ailments of the richer classes in this country. We meet every day patients who have never complained of gout, but who in point of fact are seldom or never free from that disease. These persons are, for the most part, men of enfeebled constitution, of a dark sallow, cachectic complexion, with a rough, dry skin, an irregular, often feeble and compressible pulse, scanty excretions, variable appetite, complaining of frequent heartburn and uneasiness in the right hypochondrium, of irritable disposition and despondent temperament. Such individuals are subject to a vast number of anomalous complaints, which, were they not recognized as the consequences of this diathesis, and treated accordingly, might lead to fatal results. But which, when treated by the proper remedies for gout, are most amenable to them. In this gouty habit of body mineral waters offer a most appropriate remedy.

In every age satirists have found a subject for ridicule in the multiplicity and uselessness of the remedies employed for the treatment of gout. Lucian in his *Tragopodagra* enumerates sixty substances then prescribed for the gout; of which list baths and mineral waters are the only agents mentioned by him which are still used. Even the great Sydenham chiefly relied in the treatment of his gouty patients, upon a recipe containing two and thirty ingredients, which would now sorely puzzle a modern apothecary to compound. Ovid emphatically asserts that—

"*Tollere nodosam nescit medicina podagram.*"

Nor was Fenton, who died himself from this disease, more complimentary to the power of physic, when in his "Ode to the Gout," he addresses it as—

"Thou that dost Æsculapius deride,
And o'er his gally-pots in triumph ride."

Violent remedies seldom do good in regular gout; cathartics are liable to cause recession of the inflammation from the extremities to the stomach or bowels, bleeding is now so seldom employed, even when it is required, that I need not speak of it in gout where it is not, and cold applications to the affected part are *never* justifiable. Sydenham very tersely sums up all the remedies used for gout in his day, and though a hundred and eighty years have elapsed since the opinion is still apposite. He says—" In gout, too, but three methods have been proposed for the ejection of the *causa continens*—bleeding, purging, sweating. Now none of these succeed*" Perhaps the greatest physician that succeeded Sydenham was Cullen, and though nothing could be more widely different than their views of the pathology of gout, yet how nearly do their opinions as to its treatment coincide. " The common practice," observes Cullen, " of committing the person to patience and flannel alone, is established on the best foundation."†

But during an acute paroxysm of gout, it is seldom easy for the sufferer to content himself with flannel and patience ; and fortunately, a great deal may be done in most cases in the way of mitigating the pain, and expediting the recovery. The best local application that I know of, is French wadding, with which the part should be lightly and completely enveloped, and then wrapped in oiled silk. This acts like a light poultice, when the weight of a poultice could not be borne. A draught containing a little hyoscyamus and æther, may, in most cases, be given at the same time, and followed next morning by some mild aperient, and Gregories powder, is as good a form as can be selected.

The foregoing plan of treatment, which answers for many cases of acute gout, is by no means, however,

* Works of Sydenham, translated by Dr. Latham, vol. ii., p. 131 London, 1850.

† First lines of the Practice of Physic. Aporism, DLXIX.

suited for all. For, every individual case of gout is distinguished by peculiarities, depending on the patient's age, temperament, idiosyncracy, and previous state of health; all of which must be considered in the treatment. And, herein lies the skill of the physician. The charlatan treats all cases alike, and, as an old writer well expresses it,—" sometimes he kills the disease; but more frequently, he kills the patient." The physician, on the contrary, is guided by all the circumstances of each case, and frames his treatment in accordance with them. I need hardly say, therefore, that I have not written these views of the treatment of gout with any idea of counselling a sufferer to treat himself, for no man whatever could do so with safety; but merely as a necessary portion of this general introductory account of the disease.

During the intervals of gout, medical treatment is not less necessary, and is much more efficacious than during the fit. In this treatment of the intervals of gout, I believe that modern practitioners have occasionally overlooked or despised some remedies, which, though antiquated, are still useful; as, for instance, powdered sulphur recommended by Cheyne as a " most powerful remedy in the intervals of gout. I have known," adds that eminent physician, " half-a-drachm taken regularly twice a day in a spoonful of milk, prevent the fit for many years, and lessen both its pain and duration when it happened."

About the use of colchicum in gout, it would be useless to speak in a work of this kind. In the hands of a physician, properly administered, colchicum is the most valuable medicine we possess for gout; but misused, it is one of the most injurious drugs that can be taken by the gouty patient.

As medicine is confessedly insufficient for the cure of gout, how shall the disease then be treated? Let the father of English physic answer this question. "We must," says Sydenham, "look beyond medicine... For years together a man has drunk and feasted—has omitted his usual exercise—has grown slow and sluggish—has been over studious, or over anxious—in short, has gone wrong in some important point of life."* Would it not be vain then to expect that physic could suddenly undo the work of years? As the morbid change has been produced slowly, equally gradual must be the action of our remedies. Instead of treating the gout, we must content ourselves with treating the constitutional condition of which gout is the consequence.

Gout always leaves more or less debility, in some cases amounting to utter prostration of strength, after an acute attack. Tonics are obviously indicated, and when the digestive powers are, as they generally are, impaired, I have seen more benefit from (Huxham's) compound tincture of bark than from any other preparation. When that has been useful, after a little time iron should be given, and the best forms of that

---

* Sydenham on Gout, Sections 41 and 42.

remedy are those prepared in nature's laboratory, and dispensed at the various Chalybeate Spas.

We now come to the treatment of gout by mineral waters. It is obvious, however, that this treatment can only be used as a prophylactic or preventive of the disease, or in the intervals between the paroxysms, for patients suffering from a fit of the gout could very seldom have recourse to this remedy in the only effectual way, *i. e.*, at the Spas.

Mineral waters are useful in gout in two ways. If they entirely succeed, they so change the blood by eliminating from the system principles which should not exist in it, or which exist in undue quantity, that the morbid constitutional condition which occasions gout is altered, and the return of the disease is obviated. Secondly, in cases where this entire change cannot be thus effected, mineral waters may still prove most beneficial by their action on the general health, which they improve, giving strength to the system, to localize and develope regular gout, in the stead of that misplaced, atonic, irregular, wandering form of gout now so prevalent.

By the proper use of mineral waters we open avenues through the excretory organs, for the elimination of those principles, whose presence in the blood gives rise to the phenomena of gout. The secretions from the intestinal mucous membrane, kidneys, and skin, are all augmented; and through these channels the gout-producing lithates are washed away; while, by warm bathing at the same time, the skin in these cases generally hard and dry, is softened and relaxed, thus conducing to the same end.

When the constitution is weakened, by repeated seizures of gout, chalybeates are often required; and in such cases the superiority of nature's pharmacy to that of art is shown; for often times when every preparation of steel has been tried, without benefit, some of the mild chalybeate springs "work wonders," restoring tone and strength: and thus it is that Schwalbach, Spa, or Tunbridge Wells, prove useful in gout.

Carbonated springs, or mineral sources containing carbonate of soda dissolved in water charged with carbonic acid gas, are those most commonly prescribed in the intervals of regular gout, particularly when connected with gouty dyspepsia. They correct the unnatural acidity of gouty blood, rendering it and the urine alkaline, facilitate digestion, and increase the secretions. The principal waters of this class are, Vichy, Fachingen, and Geilnau.

The simple saline, or as they are sometimes called "muriated," mineral springs, whose chief mineral ingredient is chloride of sodium, or common salt, are largely employed in gouty cases, and especially in gouty dyspepsia. They stimulate the appetite and digestion, and thus improve the blood. Their effects vary according to their temperature, some of them being cold and others thermal. Homberg, Cheltenham, and Kissengen, belong to the former, and Wiesbaden, Baden-Baden, Soden, and Cannstadt are examples of the latter.

The saline alkaline springs, which contain the soluble salts of soda, viz., the sulphate, chloride, and bi-carbonate, are principally useful in cases of gout marked by a full plethoric habit, and when intestinal, rather than gastric derangement is present. They principally act by purifying the blood from the lithic acid by increasing the renal and, above all, the alvine excretions, and secondly, by their improving the appetite, and so leading to the formation of healthy blood. The principal waters of this class are Carlsbad, Marienbad, and Franzensbad in Bohemia.

Of the English Spas resorted to by gouty sufferers, perhaps the most generally applicable are the warm saline waters of Bath; next rank those of Buxton, then the chalybeates of Cheltenham and Tunbridge Wells.

Whatever may be the mineral water selected in the treatment of any case of gout, it will seem almost superfluous to say that its curative influence must be aided by the same abstemious and guarded mode of living which the patient would have required had he been treated at home. Obvious as this seems, yet I have so often seen gouty invalids at the Spas, indulging themselves in a diet not less improper in quality than in quantity, that I feel bound to warn such patients that they cannot derive benefit from any mineral water, nor, indeed, from any other remedy, unless they bear in mind and act upon the following maxims.—I have already stated that regular gout is generally connected with a plethoric condition of body, the blood being unnaturally rich in fibrine, besides which it contains certain saline and other constituents not found in healthy blood. This state of the blood is the result of the habitual use of too much animal food, such for instance as that diet recommended by Mr. Banting for the cure of obesity. Therefore, in order to correct the faulty state of the vital fluid, the gouty patient must be abstemious with respect to animal food, should moreover refrain, as far as his general health will permit, from alcoholic liquors, which are also referable to the class of nitrogenous or azotized substances. In a word, regular gout occurring for the first time in a person of an otherwise healthy constitution, should be treated by abstinence and exercise.

## CHAPTER III.

LIFE AT THE GERMAN SPAS.

Comparison of German Spas with English Watering-places—Modes of Travelling in Germany—The Railways - Hotels—Dread of fresh Air and Water—Remarks on daily Cold Baths—Heterodox Notion —German Living. Table d'hote Dinner—Rhine Wines—Voracious Appetite of Natives –How to diagnose a German—The Cursaals— Gambling; its universal Prevalence—Natural Divisions of German Society—Medical Effects of Gambling—Its Physiological and Pathological Results.

UNDER this heading, I shall attempt to sketch my general impressions of *Vaterland*, as well as the influence of its peculiar customs and modes of life on foreign valetudinarians. For the German watering places are an epitome of Germany, seen, it is true, in its holiday aspect, but still fairly representing every variety of Germanic life and manners—royal, republican, aristocratic, plebeian, military, philosophical, religious, and black-leg. In England we have no equivalent for the German Spa. Our inland watering-places frequented by a few hundred invalids and idlers, have hardly any point in common with their German rivals, except that in both there are mineral springs.

The first thing for the traveller to think of is the

mode of locomotion in the country he visits; and as the great majority of German tourists' only experience of this is from the railroads, I shall here make a few remarks on the peculiarities of these in Germany. I shall, however, take the opportunity, when treating of the various remoter localities frequented by invalids, to describe my experience of some other modes of German travel, in the steamers on the Danube, the post-chaise in Bohemia, the *eilwagen*, or diligence, through the distant Austrian Spas, and lastly, the most enjoyable of all ways of journeying, the pedestrian, through the highlands of Saxony and the Tyrol, and the mazes of the Black Forest.

In Germany the railway fares are much lower, and the carriages more comfortable than in any other part of Europe. This is especially the case on the Nassau lines, on which the second class carriages are superior to the first on any English line, and consequently very few but English " gents " on a fortnight's excursion, or French bagmen travel in the first class in that part of Germany, excepting during a night's journey, when it is certainly preferable. The railway officials are in general remarkably civil and obliging, except in Saxony and Prussia, where they are the reverse. In the latter country there is a continual and vexatious demand made at almost every station for the railway ticket, and once I saw an unfortunate Cockney tourist, who, on being roused from a sound sleep and ordered to produce his ticket for at least the twentieth time that night, and very absurdly refused to do so, dragged, with no gentle hand from the train in which the rest of his party and his luggage remained, and left behind. But, for a happy combination of insolence and laziness, the Saxon railway employès excel any class of public or private servants in any country. These Saxon railway guards are very handsomely clad in a uniform considerably better than that of their military officers; nearly all of them wear spectacles, and their duties are apparently confined to smoking in the first-class carriages while *en route*, and drinking beer, or making love to the Hebes who minister behind the bars of the *restaurants* at the stations; meanwhile the passenger is left to look after his own luggage as he best can.

The low fares and density of the population cause immense passenger traffic. I was once forty-eight hours crossing Germany, by railway, changing from train to train, and during that long time, though there was a constant succession of travellers getting in and out at each station, there was hardly a single seat vacant for a moment in the carriage. This, however, may be in some degree the result of the absurd regulation, that no one is allowed to enter an empty compartment as long as a seat remains unoccupied in any partially filled carriage, so that sleep is out of the question in a German train.

On almost all lines of railroad in Germany there are refreshment rooms, every ten miles or so. These buffets are far better provided than the

French, and cannot, in any respect, be compared to our English refreshment-rooms, where dish-water soup, antique-looking pork-pies, and weak but boiling decoction of chicory, together with bad brandy and worse beer, are vended at only three times their ordinary prices. In Bavaria, especially, every waiting-room is a buffet, with its counter well supplied with piles of hot cutlets, German sausages, fowls, good soup, and coffee; besides which, wherever there is sufficient time, an excellent dinner, or breakfast is ready at a moderate charge, and even where the train stops only for a moment, trays with various comestibles and drinkables are carried round from carriage to carriage. As the supply must be regulated by the demand, all these signs are a proof of the importance the Germans attach to their creature comforts.

Having described the means of getting to the Spas, I should now naturally speak of the hotels in which the invalid must reside when he has arrived; but although the veritable German *gasthaus* presents noticeable peculiarities, yet as the principal hotels at all the Spas differ little, if at all, from the better class of modern French hotels, except that they are generally less expensive, I need not dwell long on the subject here. At nearly all the hotels there are two *table d'hôtes* daily, one at noon, the other at five o'clock, and almost everyone dines at either of these. The first is altogether too early, and no invalid should dine at so absurd an hour.

One peculiarity is shared by every German inn,

from the humblest *gasthaus* in Saxon Switzerland, to the "Grande" hotels of the Rhine. This consists in what may be described as an *aerophobia*, or morbid dread of fresh air. Every window is hermetically shut, and in the bed-rooms there is universally a close, musty smell of decomposing straw.

The German beds have been so often described, that I need not expatiate on these baby couches, three feet wide by five long, with their immense sloping pillows, damp sheets, and absurd eider-down coverlets, which are invariably found on the floor in the morning.

The ablutionary apparatus of a German Hotel, consisting of a pie-dish for basin, a small croft of water for a jug, a serviette used alternately for the bed-rooms and the dinner table, instead of a towel, and no soap, is certainly very uncomfortable to English travellers. But, it should be borne in mind that at the Spas, bathing in the mineral water is generally a part of the treatment, and that the necessity for the daily "tub" is done away with. Besides, heretical as the notion may seem, I have great doubts whether our almost universally prevailing custom of plunging from a hot bed straightway into a cold bath, summer and winter, is as judicious as it is generally considered.

In the first place we are told that this daily bath hardens the body and prevents cold, but we are not informed how many people are killed by the process of hardening. It is, in fact, the old story of the man who had by degrees trained his horse to live without food, but when, after infinite pains, the master had nearly succeeded, the ungrateful brute contumaciously died. And thus, I verily believe that many lives, especially of children and persons of delicate constitution and weak circulation, are sacrificed to this prevailing cold water bath mania. On the score of cleanliness, on which it is generally recommended, the cold "tub" is nearly useless, and a warm or tepid soap and water bath once a month would keep the pores more open, and leave the skin far cleaner than the daily cold bath does.

With respect to the living at German Spas, there is a wide difference in the various watering-places; in some, as at Carlsbad, the Spa doctor reigns supreme over the cook, and the hotel-keeper is obliged to submit his bill of fare each day to the physician, who arbitrarily strikes out whatever he may disapprove of. At most of the Spas, however, no such restriction is practised. The breakfasts, generally speaking, are as plain as possible—tea or coffee, with a pile of little rolls of delicious white bread, butter and honey, and, at most, eggs, are the usual morning meal. But, if thus moderate at breakfast, the Germans make ample amends for this at their "mittagsessen," or dinner, at which they are, without any exception, the largest eaters I ever saw. Though I have seen an Australian native making his first repast after a long period of enforced abstinence, and have shared the first dinner on landing of a crew which had been for fifteen long days on little more than half allowance, yet all these fall into utter insignificance before the gastronomic feats of the placid Germans, who, at mid-day, when most other people are actively pursuing their business, sit themselves down to a meal, at which they consume incredible quantities of watery soup, boiled beef with sour kreut, roast veal, uncooked smoked ham, slices of raw herring, roast goose, *kartoffel salat,* stewed fruits, and Dutch cheese, all mingled together in what appears to one unused to German living " most horrible confusion." This repast is generally accompanied by an enormous bottle of some half-fermented, acid, gout-producing Rhine wine. Of the merits of the cookery I am not sufficiently a judge to pronounce, but I may cite the rule invented by the " old man of the Brunnens," to enable the traveller to know what to expect from each course of a German dinner. " Let him," he says, " taste the dish, and if it is not sour, he may be quite certain it is greasy ; again, if it is not greasy, let him not eat thereof, for then it is sure to be sour."\*

Often have I sat by some long-haired, studious-looking, spectacled professor, whose pallid looks I had ascribed to long poring over the midnight lamp of science, but, before the dinner ended, discovered that the pale and interesting complexion was only the result of indigestion, which must follow the daily use of such quantities of pinguid meats. Two hundred and forty-nine years ago, an observant traveller made a

\**Bubbles from the Brunnens of Nassau,* p. 62, Frankfort Edition.

similar remark on the grossness of German living—
"They sit long at table," says Fynes Moryson, "and
even in the innes, as they take journies, dine very
largely; neither will they rise from dinner or supper
till, though slowly, yet fully, they have consumed all
that is set before them. . . Yea, at Berne, a citie
of Switzerland, they have a law, that in feasts they
shall not sit more than five hours at the table. And
at Basle, when the doctors and masters take their degrees, they are forbidden by a statute to sit longer at
table than from ten of the clock in the morning to sixe
in the evening; yet, when that time is past, they have
a trick to cozen this law, be it ever so indulgent to
them, for then they retire out of the public hall into
private chambers, where they are content with any
kind of meate, so it be such as provoketh drinking, in
which they have no measure, so long as they can stand
or sit. Let the Germans pardon me to speak freely,
that in my opinion they are no more excessive in eating
than drinking, save that they only protract the two
ordinary meals of each day, till they have consumed
all that is set before them, but to their drinking they
can prescribe no measure or end."*

Nor is this gross living, so quaintly described by
Fynes Moryson, confined altogether to the ruder sex.
At Baden-Baden, a very pretty "fraulein," who had

* *An Itinerary written by Fynes Moryson, Gent., first in the Latin Tongue,
and then translated by him into English, Containing his Ten Years' Travels
through the Dominions of Germany, Bohmerland, Switzerland, &c.* Part
3rd, Book 2nd, p. 83. Folio. London, 1617.

sat next to me at the *tâble d'hôte*, attracted my attention, and finding that she was so intent on the business in hand that my impertinence would not be observed, I could not refrain from admiring her gastronomic performance. The dinner consisted of fourteen dishes, and to each and everyone of these she helped herself so plentifully, that the portion she took of solid meat alone, should, in my opinion, have sufficed for a working man in robust health. I could not but sigh by anticipation, as I glanced from her, an extremely pretty, graceful girl, to her mother, who occupied the next seat, and who was a good-natured looking, very fat elderly lady, with no features to speak of, to consider that a few years of such dinners, and the nymph at my side would be metamorphosed into the counterpart of her parent.

Amid the crowd of foreigners by whom he is surrounded at the *tâble d'hôte*, the native German may be distinguished by his voracious appetite, his long, uncombed hair falling over the velvet collar of his brown coat, on which it leaves a shining margin of grease, an immense ring on the fore-finger of his dumpy, unwashed right hand, his sallow complexion, and broken discoloured teeth. For some time I attributed this latter peculiarity to continual smoking, but I soon observed that even the fair sex, who in Germany are not supposed to smoke, have the same small, stumpy, yellow teeth; and, therefore, I presume it must be caused by the acid condiments, unripe fruits, and saur kraut, of which all Germans, men and women, are so fond.

The characteristic feature of the Spas of Rhenish Germany is the *cursaal*, or gambling-house. This is generally the chief ornament of the town, and is usually an extensive palace, containing salons handsomely painted in fresco, and richly gilded and decorated. Here, reading-rooms, as luxuriously fitted-up as they could be in a West-end club, well supplied with the leading European and American journals and periodicals are found; a restaurant is at hand, where the wants of the inner man may be no less comfortably attended to; without are shady bowers and pavilions, where the choicest music, admirably performed, may be enjoyed, and all these attractions are thrown open freely to everyone. As, however, most of those who enter the salons play more or less, and as ninety-nine out of every hundred who play lose, the liberality of the *Societè des Jeux* is not altogether without its reward.

Gambling is the black spot of German life. Irrespective of the regular gambling establishments, where it is reduced to a science, and where all the ruses and shifts of the black-leg are practised with due regard to the proprieties of society, this vice infests every class of the German people, from the highest to the lowest. In every beer-shop you see respectable artizans and tradesmen assembled nightly, playing faro or rouge et noir over their beer, for groschen or kreutzers, with a steady, business-like gravity of demeanour, very different from the careless gaiety of the French "ouvrier," when engaged at dominoes or billiards for the "petit verre."

In Baden Baden, and some of the other gambling watering places society may be divided into two great classes, of sharpers and their victims, together with a sprinkling of valetudinarian spectators. The majority of the spectators are French; most of them sallow youths addicted to " le sport," dressed outrageously a l' Anglaise, whose conversation, compounded of English and French betting slang, is of the turf, turfy. Around these hover stout, over-dressed, bejewelled, close-shaved, middle-aged men, who make up for the want of a man's beard on their face by the imitation of a goat's on their double chin. A peculiarity by which these modern French "chevaliers d'industrie" may be recognised, is a certain bold, defiant air; they seem to be in a perpetual passion, speak loudly and vehemently, scowl ferociously around them; and, like Sir Mulberry Hawk in "Nicholas Nickleby," bully, as well as plunder their victims.

It is not from the habitual players that the gambling houses derive their chief profits, but from the chance visitor, who loses a few Napoleons, and then, perforce, must stop. The most profitable victim is the young tourist, who, seeing how easily a fortune might be made, adventures his five-franc piece or Napoleon at rouge et noir, and, unfortunately for himself, wins. He sees the croupier rapidly execute some cabalistic movements with the cards, hears the mysterious words *rouge perd et coulour*, and immediately, with wonderful accuracy, a second five-franc piece or Napoleon is thrown by the croupier beside his stake. Encouraged by this success, the tyro now plays more boldly, and, perhaps, wins for a time. But as the proverb has it, " Finis coronat opus ;" fickle fortune soon changes, and the luckless player may oftentimes be congratulated if he escape from the cursaal with enough money left to defray his journey home. I once met one of these unfortunate youths in Baden, where, having been so " cleaned out" that, not having enough money remaining even to send a telegram to England, he was left in pawn, as it were, for his hotel bill.

It may be said that these remarks on German gambling are out of place in a medical book on mineral waters. But, believing as I do, that this gambling exercises a most deleterious influence on the bodily health of those who indulge in it, and that it directly counteracts the curative action of the mineral springs, it seems to me the duty of a medical guide to the Spas, to point out such dangers, and to warn those who may follow his counsels, to avoid exposing themselves to such imminent peril to the *mens sana in corpore sano*. That gambling is really a dangerous pursuit for an invalid must be evident to any one who watches the play for a short time, and studies its effect on the players. In the first place the pulse is invariably quickened. This I have ascertained by direct experiments, as some of my acquaintance at Baden and Ems were good enough to allow me to examine their pulse before they entered the cursaal, and again as they left it, after playing, and I always found it accelerated; thus, sometimes, I found a pulse which was seventy-two before the play, was a hundred immediately after. The flushed face, sparkling eye, and sweating hand of a young player, are additional proofs of this vascular excitement. The respiration is also hurried, and I frequently observed novices gasping for breath, after a successful coup or a heavy loss. And, although these external signs of excitement are not observable in veteran gamblers, yet, as no training will regulate the contractions of the heart, even their pulses must be more or less excited by the gambling. Moreover, when we consider that the majority of the invalids who resort to the Spas suffer from diseases fostered by a sedentary life, and that a journey to a watering place is recommended, more because it takes away the patient from his inactive life and forces him to take air and exercise, than on account of the medicinal action of the mineral water, we must see that such a patient will not be benefitted as long as he spends his days haunting the gambling saloons of Baden, Ems, Homburg, Wiesbaden, or Spa, and his nights, bent down for hours together in a constrained position, over the roulette or rouge-et-noir table. His pulse quickened, his respiration hurried, and his mind in a state of tension, actively engaged in calculating chances, and either elated by success, or annoyed and depressed by his losses. For all men, whatever they may say to the contrary, however large their fortune or unimportant the sum at issue, are to some extent at least thus influenced by the result of their play.

Having shown that the question of gambling at the Spas is one which fairly comes within the scope of a medical treatise, I have merely to add that so impressed am I with the sanitary ill effects of this custom, that I would never advise a patient of mine to visit a Spa where public gambling was practised, without thinking it my duty to warn him, on medical grounds, of the probable evils to health of this habit; and, indeed, *cæteris paribus*, I would much prefer sending invalids to the Bohemian, Bavarian, Austrian, Swiss, or French Spas, where gambling does not exist.

How different is the daily routine of life at Carlsbad, Kissingen, or Wildbad, for instance, where gambling is not practised, to the spas at which it is? At these non-gambling watering-places, the patient rises so early that, by six or seven, a.m., he meets all his acquaintances at the promenade at the springs, where he walks about, listening to the band till he has accomplished his prescribed potations. He then, after a brisk walk in the fresh morning air, returns, with a sharpened appetite, to a light breakfast, and for a couple of hours occupies himself within doors till it is time to go to the wells again; shortly after which comes the early table d'hôte. Then excursions, long walks or drives through the neighbouring country fill up the afternoon, and he returns in time for the evening gathering at the Spa, where more water and more music await him. It seems impossible for a German to drink any mineral water, except to a musical accompaniment. The valetudinarian then retires from the night air, and closes his day at the hour which would in all probability have found him seated at dinner were he at home. Compare this regular mode of life with the excitement and confinement of Homburg or Baden, and there can be little doubt on which side are the advantages for the invalid, at least.

A chapter on German life, without allusion to beer, would be Hamlet with the part of the melancholy Dane left out. All classes and ages, and both sexes in "Vaterland" have an extraordinary, and, to me, incomprehensible avidity for swilling gallons of the weakest of small beer; for the *lager bier*, or strong Bavarian ales are comparatively little used. I have often been surprised to see the amount of "white beer" consumed by lads of fourteen or fifteen, students at Bonn or Heidelberg, without producing intoxication, but once I tasted the liquor, all wonder vanished, for so diluted and unpalatable is it, that it is more calculated to produce *emesis* than inebriety.

Smoking is certainly what our Transatlantic brethren designate an "institution" in Germany, and is there carried to a most absurd, offensive, and deleterious length. Morning, noon, and night, the immense china pipe bowl, with its unwieldy cherry stick handle, is in constant requisition. And, after my first residence in "Deutschlands," the impression left on my mind was, that every German who can afford it, spends most of his time, when not sleeping or eating, either in smoking, or else in cleaning this complicated fumigatory apparatus.

I am not one of those who believe that an after-dinner cigar does any particular harm to the majority of people. On the contrary, I think that, useless as the habit may seem, if it be, as many find it, an enjoyment, and even a solace to some of the many annoyances of life, the only reason why those who do not take pleasure themselves in smoking should condemn those who do, is that most men, as the poet has it,—

"Make amends for sins they are inclined to,
By damning those they have no mind to."

But, as abused in Germany, smoking must infallibly tend to destroy digestion, very probably leads to softening of the brain, and, no doubt, is one of the causes of the obscure mysticism in which the Germanic mind loves to envelop its ideas.

*Apropos* of smoking, I may here observe that in Lisbon I noticed, many years ago, that the male children, who up to the age of seven or eight, are generally fine, strong, healthy-looking boys, after that age very commonly undergo a striking change; their growth seems arrested, they suddenly lose the joyousness and vivacity of childhood, and become prematurely old-looking. For some time I puzzled myself to ascertain the reason of this, and it was only on my last visit, when I found that from the age of seven or eight, smoking is commonly practised by these children, that I discovered the cause.

I have, I fear, drawn no very flattering picture of German life and manners as seen at the Rhine Spas; but I would be very unjust, did I not say how different are the impressions produced by the people who live far away from these scenes of fashionable dissipation. It would be impossible to sojourn among the Southern Germans without admiring their untiring industry, simplicity of life, honesty, and kindliness of disposition.

We are used to speak of Germany as one country—whereas, its thirty-two states present the most striking difference in climate, scenery, laws, and manners, as well as in the social usages, and even disposition of the people. The cold, barren North of Germany being as different from the sunny vinelands of the Maine and Danube, as the repulsive mien and grasping disposition of the Prussians and Saxons are from the open-hearted, cheerful nature of the inhabitants of Southern Germany. The Northern Germans are certainly not a prepossessing people. They are cold and repelling in manners, dirty and offensive in their habits, prying and inquisitive in conversation, insolent and overbearing, when they may be so with impunity, but servile and obsequious when treated with firmness.

## CHAPTER IV.

### THE ART OF TRAVEL.

Its importance—English travellers—National love of travel—Its pleasures and advantages—Excuses for it—Classes of tourists—Hint to travellers—A scene at Spa—Comparisons odious abroad—Respect due to observances of others—Prejudices and luggage to be got rid of—Companions—Pedestrianism and travelling "en grand seigneur" compared.

It is important for every tourist to know how to travel with advantage, since on this much neglected art depends the whole enjoyment and comfort of the traveller. And to no class is this more necessary than to the valetudinarians who wander through the winter resorts and Spas of Europe; their health being especially susceptible to, and influenced by, whatever affects their general comfort and well-being.

A love of travelling seems inborn in the English character, and Claudianus's encomium on the superior happiness of him who, from his cradle to his tomb, had never stirred beyond the limits of his own farm—

"Felix qui patris sum transegit in arris
Ipsa domus puerum quem videt ipsa finem,"

meets few sympathisers in these islands. No people travel so widely as the English, and none travel so well under adverse circumstances. Exploring the central steppes of Asia, traversing Africa, or searching for the North Pole, our volunteer pioneers of knowledge cheerfully undergo hardships, and brave dangers that hardly any other race would so willingly face. But, paradoxical as it seems, few make such bad travellers in civilized countries. And, although our countrymen visit more foreign lands, and expend money more lavishly perhaps than any other nation, yet it is very certain that we are by no means popular abroad. Nor is it, I think, improbable that those very qualities of stubborn endurance and unbending resolution, which fit English travellers for enterprises none others would undertake, incapacitate them from accommodating themselves to foreign manners and customs.

I shall, therefore, take the liberty, as one who has watched our countrymen in the far-off settlements of Australia, and in the solitude of African deserts, as well as in almost every part of civilized Europe, of preaching a short sermon on some of the errors of travellers, and on what I regard as the art of travel.

Properly used, travelling is, I take it, the highest pleasure of which an educated and civilized man is capable, for it multiplies our enjoyments by crowding a greater number of impressions and sensations into a given space of time than anything else can do. An English traveller of the seventeenth century has left so excellent a picture of the benefits of travel that I cannot refrain from quoting his words—" But what," says Fynes Moryson, " if passengers should come to a stately pallace of a great king, were hee more happy that is led only into the kitchen, and there hath a fat messe of brewis presented to him, or rather hee who not only dines at the king's table, but also with honour is conducted through all the courts and chambers, to behold the stately building, pretious furniture, vessells of gold, and heaps of treasure and jewells? Now such and no other is the theatre of this world, in which the Almightie Maker hath manifested his unspeakable glory. He that sayles in the deepe, sees the wonders of God, and no lesse by land are these wonders daily presented to the eyes of the beholders, and since the admirable variety thereof represents to us the incomprehensible Majestie of God, no doubt we are the more happy, the more fully we contemplate the same."*

I am persuaded that at least one half of those who go as valetudinarians to the Spas of Germany, merely seek a plea, in the state of their health, for indulging their innate, natural, and most legitimate inclination for travelling. There is a fear of being thought capable of seeking enjoyment for its own sake peculiar to English professional men. The doctor, clad in a blouse, with a knapsack on his shoulder, does not like to be met by his patients crossing the Pyrenees on foot. The lawyer is not anxious to be recognized by

* An Itinerary, written by Fynes Moryson, gent, containing his ten years travells through the twelve dominions of Germany, Bohemerland, Switzerland, &c. Part 3rd, book 1, p 5. Folio. London, 1617.

his clients in a travelling jacket and with unshaven chin, scratching his name on the roof of Milan Cathedral; the clergyman, who has shaken off the restraint of his clerical garb, is not particularly gratified to encounter in the gambling salons of Baden or Homburg, some pious elder who " sits under " his pulpit in London. But once introduce the plea of delicate health, and talk of change of air, and mineral waters, and all such difficulties vanish. Though it certainly seems unaccountable why anyone should feel reluctant to own, that he travels for the pleasure, as well as the benefit of the moral and physical exercise of travelling.

Besides our invalids, real and imaginary, almost every class of English society that can afford to do so, disperse abroad each summer in pursuit of health, as well as of recreation. Our wearied legislators, exhausted by debates and committees, relax their spirits by rambling through transatlantic regions, or transalpine countries, in search of a grievance for next session. Our fagged professional men, jaded by pouring over briefs, or by attendance on disease, explore Norwegian fiords, or climb Alpine peaks. Our government employés and clerks fling off the restraint of the desk for three or four weeks, give up shaving, and, clad in the outrè costume by which English tourists may be recognized anywhere, to use their own phrase, " do " the Rhine or Switzerland. So deeply is this migratory instinct implanted, that even the British " paterfamilias," who cannot well go abroad with his retinue of children, impelled by uxorious promptings, quits his dearly prized comforts for a time, and dwells in fashion and discomfort in some terrace or dingy lodging-houses at Brighton or Ramsgate. To all these classes, except the last, it is of importance, as I have before remarked, to know how to travel; and this art is made up of particulars, which, according to the degree in which they are attended to or neglected, constitute the whole difference between an agreeable and an unpleasant journey.

In the first place, it will save the traveller much useless annoyance if he makes up his mind to the fact that as long as he travels he will be more or less imposed upon in petty matters; and, that as this is inevitable, it may as well be quietly submitted to; the only result of resistance being a loss of time and of temper, as well as of money. Besides, these continual squabbles about slight overcharges, which I have known some travellers to keep up during an entire journey, must effectually destroy the pleasure of the tour. And, after all, life is not so very long that we can afford to waste time in useless disputes over trifles.

The following scene, which occurred last season at Spa, seems to me a sufficient illustration of the foregoing paragraph. One morning, on my way to the Pouhon source, I heard a fearful din, and approaching to learn the cause, I saw a well known professor in one of our universities standing in front of the *Bapillon*, gesticulating at an old yellow-faced woman, who, with a tumbler in each hand, stood facing him, waving her skinny arms furiously about, and pouring forth, with intense volubility, a torrent of invectives in unintelligible patois. After some time the professor retired discomfited, grumbling about robbery, rights of a British subject, and police. Meeting him at the table d'hôte the same day, I ventured to ask for some explanation of the scene I had witnessed. Hardly had I put the question than the man of science, excited by the memory of his wrongs, roared out, " I have been robbed—infamously robbed, sir! and it has been the same thing ever since I entered Belgium. They are all swindlers and cheats." This, and much more to the same effect, did the learned doctor continue to vociferate to a company, three-fourths of whom were composed of the people he was abusing, before I could learn the facts of the case. It is customary at Spa for casual drinkers to give a couple of sous to the aged naiade who hands them their glass of water, and on this occasion our professor, having no copper, gave half-a-franc, expecting to receive eight sous back. But, having pocketed the coin, the water-nymph hurried off to attend to other customers, and refused to hearken to the dulcet tones with which the professor wooed her back. Irritated by such contumelious treatment, the doctor descended into the area round the well, and taxed the short-memoried fair one with her perfidy, whereupon she persisted that he had presented her with the money. Hence this whole scene, in which an English gentleman, a scholar, and a man of science made himself supremely ridiculous for the very moderate sum of four pence.

A judicious writer, whose name has escaped me, remarks that, "the traveller should consider himself a guest when abroad, and observe the same conformity with his host's manners, as he would were he paying a visit to a private house." But, unfortunately, this good advice is practically ignored by the majority of our travelling compatriots, who indulge in incessant depreciation of all the manners and institutions of foreign lands, or in invidious comparisons with something they think much better at home. I was hardly ever in company with a number of English people at a table d'hôte on the continent, that I did not hear it said that the country we were then in was ill cultivated and worse governed, the towns were poor, and the people dirty, or that the prevailing religion was superstitious. And all these unpalatable assertions are usually put forward in the most dogmatic and offensive manner, as if incontrovertible truths, and delivered in a loud voice for the benefit of the mixed company, with a happy unconsciousness that any foreigner can possibly understand English. As the Spaniards say, "Muchas veces la lengua corta la cabeza," and these rude speeches are among the principal causes of the unpopularity of English tourists abroad.

Even if such statements were as generally true as they are false, what can be the use of endeavouring to force them on the conviction of those who will not believe them. Besides, it certainly evinces a very unhappy mental constitution if, wandering through a world of such varied beauty, that wherever we turn some new object of interest presents itself to us, we should be able to see only the worst side of things, and to find that " from Dan to Beersheba all is barren."

Of the respect due to the religious observances of foreign countries, it may seem superfluous to speak, and one might imagine that no gentleman could possibly insult the form of worship adopted by any human being. But I have been on the spot when English officers were expelled from the Grand Mosque at Cairo for gross misconduct. And I have witnessed the behaviour of British tourists in the cathedrals of Spain and Italy, where, when the whole congregation were bowed down in solemn worship, I have seen men and women, clad in the garb of gentlemen and ladies, conducting themselves in a temple of religion in a manner that could be tolerated only in the gambling rooms of the German cursaals.

If a man finds his zeal for his own faith so strong that he cannot control the manifestation of his contempt for the religious usages of others, he certainly should not travel beyond the limits of his own country. And, to say nothing of the lack of courtesy, as well as of Christian charity, shown by those who, by an irreverent or scoffing demeanour, wantonly outrage the most sacred feelings of others, it surely denotes a perversion of mind to imagine that any form of religion, which is conscientiously believed, can in itself be contemptible, and a fit subject of insult and ridicule.

. Next to his prejudices, his luggage is the great incumbrance of the traveller, and like them it should be reduced to the minimum. A small portmanteau which will fit under the seat of a railway carriage, should contain enough for any man, and it is far better to buy whatever one stands in need of abroad, than to have the trouble of carrying unnecessary articles. I well remember a journey I made from Florence to Paris in company with two ladies and their maid. It was during the last war in Italy, when all the custom house and pasport officials were particularly active. The luggage of our party consisted of an immense packing case, two carriage trunks, four portmanteaus, two corded boxes, besides a pile of carpet bags, bonnet boxes, reticules, dressing-cases, umbrellas, shawls and wrappers. And, agreeable as were both my companions, and charming as was the younger, to this day I still look back with horror on that journey. What bribery of custom-house officers! What daily battles with porters! What telegraphic messages sent after missing articles, which were constantly going off in wrong directions! What long delays at out of the way places, until the stray packages should turn up! And what sleepless nights and ill-digested dinners did my anxiety about that detested luggage not occasion me! At length we reached our destination, and as I saw my fair friends off in the mail train for Calais, I registered a solemn vow never again to travel with a lady whose baggage should not be at my absolute discretion.

I may remark that some few travellers seem to regard the custom house officers abroad as their bitter personal foes, and by sarcastic observations delivered in the "unknown tongue," which cockney tourists delight to indulge in, and by their rudeness of demeanour they touch the susceptible pride of the officials, by whom they are left to the last, and then every separate parcel and box is opened, and most minutely searched, while their fellow travellers are, perhaps, comfortably dining at their hotels, or many miles on their journey. Nor does their folly always end here, for impatient at the delay, and confiding in the persuasive powers of the "oil of palms," they sometimes publicly offer to bribe the douanier, who indignantly repudiates such *open* corruption, and by way of proving his inflexible virtue becomes doubly troublesome.

I once saw an English gentleman, who had been detained for some hours at night in the custom house at Valencia, walk up to one of the officials, who happened to be the head of the department, and dropping a five franc piece into his hand, request that his luggage might be examined at once. The indignant Don, with a furious oath, dashed the money through the open doorway into the street, and made the night horrible with his roars of rage at the stupidity of the foreigner, who could not distinguish a Castilian of pure descent—*un Castellano viejo y rancio*—from a Valencian porter. In truth, a little civility goes much farther at custom houses than even money, and I would commend the Spanish proverb as a golden rule for all young travellers on the continent. " Guard thy purse, but keep honey in thy mouth" :—*Miel en boca y guarda la bolsa.*

A pleasant companion is certainly a great enhancement of the pleasure of travel. But it is a great mistake to travel with more than one fellow tourist, for when a party of three or more set out, there is sure to be some diversity of tastes or opinions, in which, instead of the mutual concessions which must be made when only two people voyage together, the majority will rule absolutely; and the unfortunate individual who stands alone in his tastes, will either be dragged to see places and things he has no desire to visit, or else will be obliged to part company and journey on by himself.

The gregarious fashion of travelling recently introduced, in which some one offers to conduct a party of tourists to Italy, Switzerland, Jericho, or any other place, and back to London, at so much a head, must be the acmé of all the incommodities of travel, divested of all its pleasures. How any individual who has the slightest vestige of self respect can deliberately bind himself to join a caravan of utterly unknown persons for a fixed period, during which they perambulate the continent after their leader, in the fashion of a flock of sheep following the bell wether, seems to me very incomprehensible. But after all, this is a matter of taste and *chacun à son goût.*

Of all modes of travel, the pedestrian is decidedly the pleasantest for a youthful and healthy tourist. There is a degree of independence about it, which is enjoyed in no other mode of travelling; and he who, with his knapsack on his shoulder, can truly say, *omnia mea mecum porto,* experiences much the same feeling of freedom of action which gives its peculiar charm to the nomad life. In the chapters on the Pyrenean watering places, will be found my reminiscences of a long pedestrian journey through some little known parts of this vast range of mountains. From the recollection of a few such excursions, I feel convinced that, for a young man in health, there is no mode of travelling for pleasure to be compared with the pedestrian.

The man who travels *en grand seigneur,* flying through countries in the coupé of an express train, or shut up in his private carriage, attended by couriers and lackeys, loses the better part of the pleasure of travelling. He sees only the external features of the country through which he passes; but misses that insight into the character, that knowledge of the manners, and that acquaintance with the every day life of the people, which expands the mind, and enlarges the sympathies of him who lives and mixes on terms of equality with those by whom he is surrounded—the latter finding that men in all countries are more like each other than is generally supposed by the former; being in general neither better nor worse in one country than in another, however much their degree of civilization and political institutions may differ.

## CHAPTER V.

#### THROUGH BELGIUM TO CHAUDFONTAINE.

The voyage—Ostend—A metamorphosis—Effect of the sea on French and English women—Reminiscences of Belgium.—CHAUDFONTAINE: Description of the village—Its thermal sources—The baths and the author's experience of them—Their medicinal application.

*Grace à Dieu*, " at last we are arrived," exclaimed a jubilant voice, as the steamer bumped against the pier of Ostend, and turning round, hardly could I recognize in the clean shaven, jovial-looking, rubicund speaker, the cadaverous, semi-moribund individual, who, an hour previously, had so faintly thanked me when, forced by common humanity, I had appeased his maritime agonies with a few drops of laudanum in brandy and soda water. " Ah c'est un beau pays Monsieur—un pays charmant," continued my new friend. Thinking I must be looking in the wrong direction I again scanned the horizon. Behind us extended the turbid waters of the German Ocean, while in front a low, barren, desolate, sandy coast lay before us, in the midst of which a small white-washed town, with one tall steeple, prominently stood out. Seeing, I fear, that I hardly coincided in his admiration of the country, my new friend, a cotton manufacturer from Liege, who had been on a pleasure trip to England, said, " I will fetch my wife and we shall dine together," and, descending to the cabin, he returned accompanied by his better-half and two very pretty daughters. Waiting quietly till the impatient crowd of travellers had hurried on shore, where they had to stand for half-an-hour in the rain till their luggage was debarked, we landed and made our way to the douane, where the examination was a mere form. From the custom house, a few minutes in an omnibus brought us to the Hotel d'Allemagne, opposite to the Railway Station. Wonderful as had been the transformation in the appearance of the gentleman we have described when ill at sea and recovered on landing, it was nothing when compared to the change in that of his wife and daughters, when, in half-an-hour after our arrival, in a very becoming toilet, they joined us at dinner. Indeed, it has long been a mystery to me how it happens that French, German, and other continental women, be they ever so good-looking and well dressed when they come on board ship, after being a few hours at sea invariably become tossed, and dowdy-looking. I have noticed this in many parts of the world, in voyages of all lengths, from three months to as many hours, and have been struck by the fact that neither English nor American ladies are affected in the same way. What the explanation may be I know not, except that our fair countrywomen are less dependent than any others on those adventitious aids to beauty, to which foreigners are so largely indebted.

After dinner, at the Hotel d'Allemagne, which to us hungry travellers seemed very fair, and the wines unexceptional, we strolled through the uninteresting streets of Ostend, and along the Digue crowded with fat burghers from Ghent and Brussels into the Casino, where, amid a horrid din of excited Belgians, and in an atmosphere so dense with the fumes of bad tobacco that it was barely possible to distinguish one's next neighbour, Rouge et Noir and Ecarté were being played.

Ostend is certainly not a place to remain in longer than one can help it, and, being anxious to advance on our journey, we thought it now time to return to the railroad, so bidding adieu to our Belgian acquaintances, we arrived at the station in time for the last train to Bruges.

Of the Belgian towns we visited, I shall spare my reader any account, not that English travellers, en route to the Spas, generally see all that is worth seeing in Belgium. Indeed most tourists to Rhenish Germany now see nothing whatever of Belgium. They take a through ticket from Victoria or Charing Cross to Cologne, enter a first-class carriage at either of these places, are whisked to Dover at the rate of sixty miles an hour, rush on board the steamer, and after forty minutes of misery find themselves in Calais, spend ten minutes in the refreshment room, and are again whirled away, through the darkness of night, to Cologne, where they arrive, jaded and fatigued, in nineteen hours from the time they left London. And this is called travelling for pleasure! For my own part, I must confess that I cannot relish being shot from city to city in this cannon-ball fashion of travelling, except when urged by dire necessity, and, above all, through a country like Belgium, where every town —nay, every village—presents remains of historic interest that can be found in no other part in Europe. I love to loiter through the long-deserted streets of Bruges, nor forget to admire the "formosis Bruga puellis," or to contrast with it the busy hum of commerce which still pervades the old warlike city of Ghent; or losing my way in the quaint, winding thoroughfares of Antwerp, to stray through its glorious cathedral and churches, and gaze for a time on the masterpieces of Rubens, Teniers, and Vandyk. I should however weary my readers were I to dwell longer, as I might easily, on the sad loss English tourists inflict on themselves by passing through this, the best governed and most prosperous, as well as most interesting country in Europe, without seeing anything of it, except, perhaps, Brussels—the least Belgic town, although the capital of Belgium.

The first Belgian Spa on the route from England to the Rhine is Chaudfontaine, and if facility of access, beauty of situation, and convenience of living, were the chief recommendations for a watering place, then should Chaudfontaine be among the most resorted-to spas of Europe.

Chaudfontaine is a thermal mineral spring, having a temperature of 92°, and is principally used for bathing; it contains but two grains and a half of saline matter to the pint of water. Of this more than one-half is common table salt, and the rest is chiefly carbonate of lime.

Chaudfontaine, which lies five miles from Liege, in the valley of the Vesdre, is enclosed by beautifully wooded hills, dotted here and there with the country villas of the Liégois. The valley itself is of that bright fresh green hue, so rarely seen on the continent. From the railway station, a little bridge across the Vesdre, leads to the village, which is situated immediately at the foot of the hills. It consists of a long straggling street, upwards of half-a-mile in length, and contains between forty and fifty houses, nearly all of them being hotels, cafès, or lodging houses. Immediately opposite the bridge, is the "Hotel des Bains," in the court yard of which the thermal spring rises. The place bears the look of a quiet country village; the hotels are small and unpretending, there are no shops, and there is none of that attempt at fine architecture, which all the German watering places aspire to. Near the railway station, between it and the river, a pretty bright looking Cursaal has been erected, and contains a handsome central saloon, where concerts and balls are given; a cafè—

restaurant, billiard and reading rooms, well supplied with English, Russian, French, and German papers, and to this building, as well as to the gardens about, all respectably dressed persons are admitted free of charge, except on Thursdays, when an admission fee of twenty-five centimes is demanded. The Bath house is a plain commodious building, lying between the Hotel des Bains and the Hotel d'Angleterre, and contains about thirty-two baths. The rooms are small, and very simply furnished, but well ventilated.

Having in vain endeavoured to persuade my travelling companion to try the effects of a mineral bath, I was perforce reduced to the necessity of trying it myself. The temperature of the water, as it flowed into the bath, was 92°; it is clear and limpid, and produced a somewhat "soapy" sensation. I certainly did not experience the remarkable feelings of *bien être* that others have attributed to this bath, nor did I notice, as they have, a greater whiteness of the skin while in the bath than is usual in any other warm bath.

Dr. Lee says that the quantity of water supplied is so great, that one may leave the overflow valve open, and thus secure a constant current of the same temperature. Accordingly I did so, but in a few moments found myself stranded in the empty bath, like a fish out of water, the escape pipe being much larger than that for the entrance of water.

One great defect in these bath rooms is, that they become filled with steam, which condensing on the walls, runs down in a stream on the bathers' garments.

The season commences on the first of May, but the principal month here is July, and during last July, the manager of the thermal establishment informed me that no less than 2543 baths were administered.

The price of the baths in Chaudfontaine, as compared with the German and French baths, is extremely moderate; thus—

| | | | | |
|---|---|---|---|---|
| A simple bath | costs | | 75 centimes | |
| "   " with linen | " | 1 franc | | |
| "   " first class | " | 1 " | 50 | " |
| A dozen tickets, ditto | " | 15 " | | |
| A cold bath | " | 1 " | 50 | " |
| Douch | " | 2 " | | |

According to the directions hung up in the rooms, the duration of the bath should be one hour: but this, I believe, is too long. I myself only remained half that time, and even that I think is too much, for the whole evening after my first bath, I felt feverish, uncomfortable, and not so well as I had been previously; I therefore think that from fifteen to twenty minutes are sufficient for a beginner to remain in the bath.

The physiological effects of the Chaudfontaine water are generally said to be calming or sedative; and it is principally resorted to in cases of nervous irritation, neuralgia, chronic rheumatism, contractions from wounds, or following long continued rheumatism. I should myself, from what I have seen of their effects, be inclined to recommend the baths in some cases of female ailments, depending on defective functional discharges. The internal use of this thermal spring is also resorted to, not as an independent remedy, but simply as an adjuvant to the external use of the water. The baths of Chaudfontaine are counter-indicated or forbidden in all cases where a disposition to hemorrage, especially from the lungs, exists, and where there is any tendency to latent inflammation, which might be thus called into active existence.

## CHAPTER VI.

### SPA.

Situation of the town—Its chief features—Population fixed and transitory—Doctors—Beauty of surrounding scenery—Advantages for valetudinarians—Hotels and living—The "Redoute" — The season—Characteristics of society here contrasted with other watering places—Analysis of the mineral springs—Detailed account of visit to each source—Action of the waters—Patients who should be sent there.

Spa possesses two great advantages; it is the most accessible, and one of the most enjoyable of the continental climate watering places, though the mineral waters are not the most powerful of this class. It may be reached in twenty-four hours by steam and rail, from London; but I should be very loathe to travel with the man who, on any but the most urgent necessity, would thus rush through the most interesting country in Europe.

The town lies in a valley, in the Ardennes hills, two of which completely shelter it from north and east winds; but still, as it stands nearly a thousand feet above the level of the sea, it is by no means exempt from cold and changeable weather. The little town itself is neat and clean, and demands little or no description, as it consists mainly of one long street, ascending the hill; this, at the foot of the acclivity, forks into two avenues, and near their intersection is the " Pouhon," or principal fountain.

In one of these thoroughfares, is the " Redoute," and a little lower in the same street, is the new bath house, the only buildings in Spa with any architectural pretensions, the other houses being small shops, hotels, and lodging houses. At the lower end of these streets commences a long boulevard, which runs to the railway station, and off this is a very beautiful avenue, the *Promenade de Sept Heures*.

On the first Boulevard, are several of the principal hotels, of which there are some twenty-four to choose from, which for the most part have the character of being comfortable and moderate.

The resident population does not exceed four thousand, whose health, together with that of the thirteen thousand visitors who annually pass the season here, is looked after by five physicians, three apothecaries, and three midwives. One of these physicians, Dr. Cutler, is an English gentleman of high professional reputation.

Spa, I remarked, is one of the most enjoyable watering places; and by this I mean, that the beauty of the surrounding scenery is such, and there are so many easy and picturesque walks and drives through the woods and hills of the Ardennes, that the invalids who come here weak and languid, are so braced up by the waters, as to be in a position to take advantage of these, and perfect their cure by their minds being diverted from themselves by the beauties of nature, and by their blood being thus circulated, and their spirits exhilarated by having recourse to nature's great restorative, open air exercise. But those who mean something very different from open air exercise by enjoyment, will find in Spa, a gambling table, open every day from morning till midnight; a theatre open on Sundays, Tuesdays, and Thursdays; a public ball on Wednesday and Saturday, and a concert on Monday and Friday. And at each of these places of amusement, those who come abroad for sanitary benefit, may inhale a highly deleterious atmosphere, and make fresh inroads on their small stock of health in the most satisfactory manner possible.

The living at the hotels here is better than at the generality of the German watering places. The small mutton of the Ardennes is equal to the Welsh, and the neighbouring streams furnish an abundant supply of excellent trout, carp, and roach.

The "Redoute," or gambling house, is situated in the main street, not far from the Pouhon fountain. The lower part of the building is fitted up as a café, while the upper contains the rooms for play, and here, every one is admitted whose dress indicates a satisfactory state of purse. These rooms, though handsome, are not to be compared in richness of decoration with the salons of Homburg, Wiesbaden, or Baden; nor is the gambling on the same scale, the usual stake being a two-franc-piece. The play, too, seemed to me to engage less of the attention of the spectators, there was little of the feverish anxiety so commonly witnessed at other cursaals, and the people lounged about in groups, with little appearance of interest in the gaming. One old gentleman with a patriarchal white beard, and a most benevolent expression of countenance, however, contrived to win upwards of ten thousand francs in the half-hour we were looking on.

The season commences on the First of May, and ends on the last day of October, but hardly any one remains after the end of September, as the weather then generally becomes cold and broken. The mineral springs of Spa are said to have been discovered in 1326, and from the works of De Steers, we gather, that in the early part of the seventeenth century they were much frequented by English invalids.

Spa was formerly the most aristocratic watering-place in Europe—from the time that Henry the Third of France patronized it in the sixteenth century, it seems to have attracted all the crowned heads of Europe in succession, and in 1783 no less than thirty-three royal highnesses, princes, and dukes, drank these waters. Now hardly one royal personage visits Spa during the season, but yet it keeps up a kind of aristocracy of its own. Considering the large number of visitors, amounting last season to upwards of thirteen thousand, there is surprisingly little of the vulgar ostentation, fast life, and society of the black-leg, and demi-monde character, which is so obtrusively forced upon one's notice in several of the spas of Rhenish Germany. Gambling exists here, it is unfortunately true; but it is only as an episode, and not the sole business of the place, as in Baden-Baden and Homburg.

Spa belongs to the class of acidulous chalybeate cold mineral waters. But as the fountains present some difference in their chemical constituents, which, though slight, is yet sufficient to modify sensibly their medicinal properties, it will be necessary to examine each of the sources separately, and then point out the general application of the waters.

ANALYSIS OF SEVERAL MINERAL SPRINGS OF SPA.—
CONTENTS OF SIXTEEN FLUID OUNCES.

| Sources. | Temperature. | Carbonic Acid Gas (in cubic inches). | Solid Ingredients (grains). | Carbonate of Prot-oxide of Iron. | Carbonate of Lime. | Chloride of Sodium. | Carbonate of Soda. | Carbonate of Magnesia. | Silica. |
|---|---|---|---|---|---|---|---|---|---|
| Pouhon | 50° | 21 | 3·37 | 0·87 | 0·75 | 0·20 | 0·90 | 0·31 | 0·23 |
| Geronstére | 49· | 14 | 1·65 | 0·45 | 0·33 | 0·09 | 0·45 | 0·16 | 0·10 |
| Souvenire | 49· | 20 | 1·28 | 0·43 | 0·22 | 0·06 | 0·30 | 0·10 | 0·07 |
| 1st Tonnelet | 51· | 22 | 0·96 | 0·39 | 0·15 | 0·04 | 0·21 | 0·08 | 0·04 |
| 2nd Tonnelet | 51· | 19 | 0·58 | 0·25 | 0·12 | 0·01 | 0·08 | 0·03 | 0·02 |
| Groesbeck | 50· | 21 | 0·83 | 0·24 | 0·16 | 0·04 | 0·22 | 0·08 | 0·04 |

The chief spring is the Pouhon, which is situated nearly in the centre of the town. It rises in a marble basin, about three feet deep, and is surrounded by a paved area, some ten feet under the level of the street. Round this each morning congregate the valetudinarians, while the women who fill the glasses bustle about the source, dealing out the water in exchange for sous. The spring rises out of a soft ferruginous slaty rock, and is of a bright clear colour; although the chalybeate nature of the water is indicated by the reddish coating of protoxide of iron deposited round the sides and bottom of the well. The taste is more decidedly ferruginous than that of any other of the springs of Spa. It is acidulous and piquant however, and by no means unpleasant, owing to the large amount of carbonic acid gas it contains. The gas is not so apparent to the eye in this water as it is in the *Tonnelets*, but in a moment or two after being poured into a glass is disengaged, and adheres to the side of the vessel in minute bubbles, to such an extent as to render the glass perfectly opaque. Immediately above and behind the fountain is, what is called the "Temple de Pouhon," an ugly room with a colonnade in front, which affords shelter to the water drinkers in wet weather. The wall of this edifice is decorated with a pompous inscription to the memory of that illustrious savage, Peter the Great, who graciously condescended to recover his health here.

The Pouhon is the mineral spring *par excellence* of Spa. It is the richest in iron of all the sources, and contains twice as much other saline ingredients as any of the rest. Moreover it is the only one of the Spa waters that is exported.

The Pouhon is especially recommended by Dr. Sutro, " in obstructed portal circulation, in deficient bilification, in congested liver and spleen, following intermittent fever; also in flatulency, digestive weakness and acidity, tendency to diarrhœa and passive hæmorrhage."* It is also occasionally employed in cases of dropsy supervening on acute diseases, and it was in this malady that Peter the Great found the Pouhon so efficacious.

On our pilgrimage to these springs, we first directed our steps to the latest discovered source, viz., the *Barisart*, which lies about a mile and a half outside the town.

The spring rises in a kind of rocky grotto, and was administered to us by a young English girl, whose family own the adjoining restaurant, which seemed to attract more customers than the fountain. The temperature of the water is 52°. It is clear and sparkling, and evidently contains more carbonic acid gas than the Pouhon; its taste too, is decidedly stronger and more ferruginous than that source—although it really contains much less iron. I found it a matter of some difficulty to take the specific gravity of this, as well as the other springs of Spa, as they contain much free carbonic acid which adhering to the bulb of the hydrometer might occasion error on the point; thus half a minute after immersion the instru-

* Lectures on the German Mineral Waters, by Sigismund Sutro, M.D, p 327, London, 1851.

ment floated at 1005, and in five minutes it rose to 1032.

The Barisart spring is said to agree in many cases in which the Pouhon is found too powerful, and is held to be especially useful in cases of atonic dyspepsia, excessive debility after convalescence from fever, and in hepatic obstruction. In one case which fell under my observation an elderly gentleman recovering from disease of the kidneys by which he was so weakened as not to be able to walk when he arrived in Spa, tried the Pouhon for some time without deriving much advantage, and being persuaded to change it for the Barisart did so, and in a few weeks was so much stronger as to be able to walk out to Barisart, drink ten or twelve ounces of water, and walk back to Spa before breakfast.

From Barisart a very pretty walk of half a league, through a thicket, along a noisy brook, the rocks of which showed evident traces of iron, brought us to the Geronstere source, which lies about three miles south of the town, in the midst of a thick wood. This spring rises under a very pretty open pavilion, and may be discovered before it is seen by the strong odour of sulphuretted hydrogen evolved from the water, which is clear but not sparkling like the other springs, and in fact contains less carbonic acid than any of them. It is much weaker than the Pouhon, and is used in cases in which an acidulated weak chalybeate water containing sulphuretted hydrogen is indicated, as preparatory to the strong sources.

Leaving the last fountain, a walk of three-quarters of a league by the road to Malmedy, conducted us to the springs of Sauveniere and Groesbeck, which are situated about a mile and a half to the south-east of Spa. The first of these sources is contained in the court-yard of a neighbouring inn, enclosed under a wooden shed. The sides of the well are deeply stained by an incrustation of protoxide of iron, of which this spring contains about as much the Geronstere, but differs from it in being very sparkling, as it possesses nearly as much carbonic gas as the Pouhon. The taste is saline, ferruginous and acidulated. Close by the side of the well is the form of a shoe deeply engraved in the rock, and tradition asserts, and ladies believe, that whoever quaffs a glass of the Sauveniere standing, with her right foot in the "pied de St. Remacle," will within the year augment the population.

A few yards from the Sauveniere is the Groesbeck spring, so named from some Baron de Goesbeck, who, very wisely fore-seeing that posterity might otherwise ignore his existence, in 1771 put a pompous inscription over this well, commenting on its virtues and his own. The taste of this source is quite different from the Sauveniere, from which it is hardly five yards distant. It is much less ferruginous and saline, but somewhat more gaseous than the last described well; containing, indeed, less iron than any of the springs, or not quite a quarter of a grain to the pint of water.

At a little distance from these springs, in the adjoining wood, is a monument erected by the Duchess of Orleans, in gratitude for her restoration to health here in 1787. This was destroyed at the time of the French revolution, and was rebuilt by Louis Phillippe, in 1841.

Our old guide, who at first had watched our proceedings with surprise when I tried the temperature and specific gravity of the springs, got tired of the repetition of the same performance at every source, and here remonstrated against the delay it occasioned. Accordingly I deemed it judicious to appease his wrath with beer, *à discretion*, which he swallowed with infinite gusto, and was so mollified thereby as to allow us to pursue our experiments in peace. Though, from the suspicious side-glances he gave from time to time, I imagine he had concluded that my companion and myself were endeavouring to poison the springs; but after the episode of the beer he evidently took a more favourable view of the matter, regarding us as only harmless lunatics, and even consented to carry my thermometer and case of tests.

We turned off from the high road at this point, and followed a narrow pathway, through the forest, which contains a few fine trees, but is impassable, excepting by the paths, from the thickness of the underwood. I never saw anything to equal the profusion and variety of the fungi, and while stooping down to examine a mushroom that was new to me, a small snake, about eighteen inches long, beautifully variegated, glided out from the cover and almost passed over my hand. My disappointment at not being able to secure it was considerably lessened when the guide informed me that it was of a well-known venomous species, and detailed several instances in proof of the assertion. The woods about Spa, I afterwards learned, abound in several species of these reptiles, which are especially troublesome during winter. From this foot-path we again entered on the high road, and, passing by a small village, arrived at the Tonnelets.

These springs were so named from the little wooden vessels in which the water was originally collected. There are three Tonnelets, two of which are under cover, and the other rises in the open air. Tonnelet "No. 1" is the source contained under the shed facing the high road. The sides of the well are deeply incrusted with protoxide of iron. The temperature of the water is 52 deg.; it is very bright and sparkling, and the taste is acidulous, sharp, and agreeable. "No. 2" is situated at the opposite side of the same building; in most respects it resembles "No. 1," but its taste is somewhat sulphurous, and the bubbles of carbonic acid gas are larger, and in greater abundance, breaking on the surface at the end of every four minutes, during which interval the water remains quiescent.

"No. 3," the original Tonnelet, rises outside and to the right of the other springs; unlike them, it is uncovered, and appears neglected. The taste is decidedly sulphurous but is not unpleasant, owing to the large amount of carbonic gas it contains. These springs I regard as the most important in Spa, although they are much less frequented than the Pouhon or Barisart. The most efficacious of the Tonnelets is that which I have described as "No 1," and to it the following observations principally refer. It is by far the most agreeable of the Spa mineral waters, and as the rest have not inaptly been compared by Dr. Wilson to Seltzer water, I think we may adopt the phraseology with regard to this source, of an old Irish writer on mineral waters, whose works, though now little known are, nevertheless, still well worth reading—namely, Dr. Lucas, who in his "Essay on Waters," published a hundred and ten years ago, in which he terms the Tonnelet "mineral champagne," says: "To the taste it is most gratefully subacid, vinous, smart, and sprightly, not unlike the briskest champagne wine, imparting exceeding little, if any, vitriolic or ferruginous taste. Being drank, it generally sits lightly and agreeably on the stomach; and, though excessively cold, it warms, cheers, and invigorates."* Besides the springs I have described, there are some others probably not less powerful as remedial agents— these are the Pouhon-Pia, about a quarter of a mile to

* *An Essay on Waters*, by C. Lucas, M.D., vol. ii., p. 199—London, 1756

the north of the Geronstere; Watroz in a valley between two rivulets, a quarter of a mile north of the Sauveniere; and lastly, the fountain of Niveset, situated at a wet, boggy plain, about a quarter of a mile to the east of the Tonnelets, and considerably lower in the valley than those springs. It would be adding unnecessarily to this chapter to give any detailed account of these latter sources as they are no longer fashionable, and therefore I need hardly add are no longer used medicinally. For what fashionable invalid could be persuaded that he might derive advantage from any mineral water, however excellent in itself, which was resorted to merely by pauper patients who only want to be cured; while at a neighbouring fountain he might actually drink out of the same well with a scorbutic Lord or a cachectic Lady.

In some cases of consumption, which I have met with, the disease could be traced back to a severe attack of scarlatina, measles, or some other febrile affection, the issue of which was doubtful, the convalescence protracted, and the recovery imperfect, leaving behind, as I have seen even in boys previously of strong and hardy frames, debilitated and ailing constitutions, in which, after perhaps a few years, consumption developed itself. In such cases I think that a few weeks at Spa, or any similar watering place, would be more likely to re-establish the child's health, and avert consumption, than any other measure.

Anæmia, in all its varied forms, in which there is a

deficiency of red blood corpuscles, is treated efficaciously by these waters. In Chlorosis, Spa is recommended to the fair sex, and ladies who wish to show their practical contempt for the theory of the Rev. Mr. Malthus, are promised a specific in the Sauveniere. Some forms of dyspepsia and irregular alvine action, whether marked by excessive irritability of the intestinal mucous membrane, or as is more commonly the case, by torpidity and constipation, both of which may be caused by debility and want of tone; when they can be clearly traced to this cause, may be treated by the chalybeate Spa waters. They act also on the kidneys, and were formerly considered to be very useful in cases of gravel.

The dose in which the Spa waters should be taken is laid down to a drachm in most of the books. We are told that formerly from two to three hundred ounces of water were often consumed daily? and that Peter the great used to dispose of sixty-three ounces, or twenty-one glasses, each morning. I doubt very strongly, however, that any ordinary mortal, not being Great, could manage this quantity.

## CHAPTER VII.

### AIX-LA-CHAPELLE.

*The journey from Spa—A breakdown, of which the author avails himself to digress—Arrival in Aix—The Dragon d'Or—Dr. Velten—Account of AACHEN—The mineral springs—Analysis—Their common origin proved—Their physiological effects and remedial power. Influence on chronic rheumatism, gout, and cutaneous affections—The Baths—Their modes of administration and action—BORCETTE: Its hot springs and their uses.*

THE distance from Spa to Aix-la-Chapelle is barely thirty miles, and yet it took us four hours of railway travelling to accomplish this short journey. A little beyond Pepinstere a goods train had broken down on the line before us, and consequently we were "shunted" for a couple of hours. And I may here remark, that none of the casualties and annoyances of travelling seem to affect the British tourist so much as this detention. He appears to resent it on principle, and to regard it as a direct violation of magna charta, and a monstrous infringement of the liberties of an English subject. I have observed the behaviour of various people under such circumstances, and have never seen any who take the matter so much to heart as our compatriots do. I have been shunted for half a day under an Egyptian sun in August, locked up in a small, ill-ventilated, well-cushioned compartment, the atmosphere of which gave one a lively impression of the oven in which Mons. Chabert, "the fire King," was wont to exhibit. And there, within half a dozen miles of our destination, El Masr, I was shut up in company with a couple of turbaned Turks, who for four long hours sat silently watching my vain efforts to effect a release, till finally one of them, a patriarchal-looking white-bearded Iman, spoke for the first time to his companion, saying, " Wallah ! the Effendi is mad ;" to which the other gravely replied, " Insh-allah ! if it please Allah he is ;" and then both relapsed into placid silence for the remainder of our detention.

I have been shunted, too, of a midwinter's night, in the Basses-Pyrenees with a party of Gascons, while the line was being cleared from the accumulated snow, and never for a moment did their sprits flag ; they laughed, and sang, and gasconaded, notwithstanding the intense cold, as cheerily as though they were comfortably at home. I have been " shunted " with a train full of German students en route for the fair of Canstadt, near Stuttgart, for a couple of hours, during which these long-haired youths ceased not for an instant the hideous yells and discordant roars which pass current for songs in German university towns. But of all the people with whom I have been " shunted," most extraordinary is the effect of this operation on the English tourist. For, almost invariably I have remarked, that the general indignation which is the result of this process is so great as to break down the otherwise impregnable and impenetrable barrier of national reserve; and the fellow sufferers league together for the moment in terrific denunciations of vengeance, and threats of actions for detention and false imprisonment. And thus it was on the present occasion; but as no railway official took the least notice of us, and no guard or porter presented himself on whose devoted head we might wreak our united wrath, our amity gradually wore away, and we all relapsed into our normal, dignified, mutually repellent and defiant attitude, and mused on our wrongs in moody silence, till in due time, the line being cleared, we got under way; and, changing carriages at Verviers, finally arrived at midnight in AIX-LA-CHAPELLE.

There is no lack of good hotels in Aix-la-Chapelle and the " Dragon d'Or," in the Comphausbad Strasse, where we took up our quarters, is probably as good as any of them; it has an excellent cuisine, but is less frequented by English visitors.

We had omitted to order ourselves to be called next morning, and so, wearied by our journey, slept on undisturbed till noon, when on descending we found our breakfast laid on the table, at which some forty guests were finishing the early table d'hôte dinner. For a moment every eye was fixed on us as we entered, and methought I detected a look of virtuous indignation at the lateness of our morning repast, in the prolonged stare of a stout lady, who with marvellous celerity had just emptied a large glass of kirch wasser.

"You make a great mistake, sir," said my next neighbour, an old Dutch gentleman, as he informed me, for his English bore the smallest trace of a foreign accent. "You make a great mistake if you come to Aachen for your health," he continued, "to begin the day at an hour when the inhabitants have half finished theirs. Every one here rises at five or six o'clock in the morning, goes to the baths, makes a light breakfast, then takes his walk, come home to dinner at noon, rests for a couple of hours, again visits the springs, has perhaps a light supper, and spends an hour or so in the cursaal, and returns to bed by ten o'clock. Try it for a day or two; and besides, the early dinner is generally much better than the late in Germany." I thanked my mentor for his friendly counsel, and must say that whenever I tried it I always found that his observation about the early dinner being better, although often not half the price of the later table d'hôte, was perfectly correct.

Our lecture and our breakfast ended, we went out to present some letters of introduction, and first called on Dr. Velten, the principal physician of Aix-la-Chapelle, who very courteously accompanied us to every thing worth seeing in the town, and especially its mineral springs, baths, and medical establishments; and to his assistance I owe much of the information I have collected on these subjects.

## AIX-LA-CHAPELLE.

Before entering into any details on the mineral and thermal waters of Aachen, I will say a few words on the general topography of the place.

Interesting as are the reminiscences attaching to this, perhaps, the most historic city in Europe, yet I must necessarily leave out any allusion to these in a work of this nature, and a very brief notice will suffice for the town as a modern watering place.

Aix-la-Chapelle, or Aachen, as the Germans love to call it, is situated some thirty miles to the south of Spa, in a fertile valley at the foot of a range of well-wooded hills of no great height. Like most of the watering places, Aachen consists of an old town immediately around the wells, and a modern suburb, in which the invalid visitors pitch their quarters as far away as the extent of the place will allow of, from the waters they have come to make use of. This latter portion occupies the upper part of the town on the Borcette or south side, extending from the theatre to the railway station; and in it the streets are wide, well paved, clean, rectangular, very dull and uninteresting.

The springs are divided into upper and lower, of which the former are the hottest. The "Elisenbrunnen," which is the principal mineral drinking fountain, is supplied by the "Kaiserbad," the warmest and strongest of the sources, whence the water is conveyed to its basin by pipes. This fountain is situated midway between the old and new towns, close to the theatre, and issues some fifteen feet below the surface, under a handsome dome, in the centre of a long colonnade facing the street, and here the water drinkers promenade, whatever may be the weather, from 5 to 8 a.m., and again in the afternoon. The water is perfectly clear, but, to an unaccustomed palate, at least, is horribly nauseous and strongly odorous of sulphuretted hydrogen.

I am indebted to my friend Dr. Velten for the following tables, which exhibit the composition of these waters according to Libeg's last analysis, and show, that in addition to the substances previously found in them, they contain iron, potassa, iodine, and bromine.

COMPOSTION OF THE AIX-LA-CHAPELLE SULPHUR WATERS.

(A.) In one thousand parts:

| Not volatile ingredients. | Emperor's Spring | Cornelius Spring. | Rose Spring | Quirinus Spring. |
|---|---|---|---|---|
| a. In ponderable quantity. | | | | |
| Chloride of Sodium | 2·63940 | 2·46510 | 2·54588 | 2·59595 |
| Bromide of Sodium | 0·00860 | 0·00360 | 0·00360 | 0·00360 |
| Iodide of Sodium | 0·00051 | 0·00048 | 0·00049 | 0·00051 |
| Sulphuret of Sodium | 0·00950 | 0·00544 | 0·00747 | 0·00234 |
| Corbonate of Soda | 0·65040 | 0·49701 | 0·52926 | 0·55267 |
| Sulphate of Soda | 0·28272 | 0·28664 | 0·28225 | 0·29202 |
| Sulphate of Potassa | 0·15445 | 0·15663 | 0·15400 | 0·15160 |
| Carbonate of Lime | 0·15851 | 0·13178 | 0·18394 | 0·17180 |
| Carbonate of Magnesia | 0·05147 | 0·02493 | 0·02652 | 0·03346 |
| Carbonate of Iron | 0·00955 | 0·00597 | 0·00597 | 0·00525 |
| Silica | 0·06611 | 0·05971 | 0·05930 | 0·06204 |
| Organic substance | 0·07517 | 0·09279 | 0·09151 | 0·09783 |
| Carbonate of Lithia | 0·00029 | 0·00029 | 0·00029 | 0·00029 |
| Carbonate of Strontian | 0·00022 | 0·00019 | 0·00027 | 0·00025 |
| b. In not ponderable quantity. | | | | |
| Carbonate of Manganese | — | — | — | — |
| Phosphate of Alumina | — | — | — | — |
| Fluate of Lime | — | — | — | — |
| Ammonia | — | — | — | — |
| Sum of the not volatile contents | 4·10190 | 3·73056 | 3·89075 | 3·96968 |

## COMPOSITION OF THE AIX-LA-CHAPELLE SULPHUR WATERS.

(B.) As regards the gaseous contents, which are developed by boiling in a vacuum.

1,000 cubic-centimetres (=1 Litre) of water containing—

| Absorbed gases in cubic-centimetres. | Emperor's Spring | Cornelius Spring | Rose Spring. | Quirinus Spring. |
|---|---|---|---|---|
| Azote | 12.78 | 12·54 | 14·71 | 7·31 |
| Carbonic acid | 126·94 | 148·46 | 145·40 | 106·30 |
| Bihydroguret of carbon | 0·52 | Slight. | 0·89 | 0·30 |
| Sulphuretted hydrogen | — | — | — | — |
| Oxygen | 1·76 | — | — | 0·09 |
| Total in cubic-centimetres | 142·00 | 161·00 | 161·00 | 114·00 |

(C.) GASES RISING IN THE WATER.

100 volumns of the Emperor's Spring contain:

| | |
|---|---|
| Azote | 66·98 |
| Carbonic acid | 30·89 |
| Bihydroguret of Carbon | 1·82 |
| Sulphuretted Hydrogen | 0·31 |
| Oxygen | 0·00 |
| | 100·00 |

Although from the preceding tables it might be imagined that the sulphuretted hydrogen contained in these springs is the least important of their gaseous constituents, yet by practical experience we find that it is by far the most active of them; the medicinal properties of the waters depending in a great measure on the amount of sulphuretted hydrogen they contain. This gas it here combined with a larger amount of nitrogen than in any other European sulphurous spa.

From the foregoing tables we see that there is no

very great difference between the composition of these springs, nor, indeed, can there be any real variation between them, as they all rise from the same source. This was recently shown, on the occasion of the Emperor's source being pumped out for the purpose of cleaning the well, when all the other mineral fountains were equally affected by this operation, thus directly proving the communication between them. The temperature of the several springs, however, varies according to their position with respect to their common source.

There is a chalybeate spa in the "Theater Strasse," with a bathing establishment attached to it, the water of which is cold, and is said to contain half-a-grain of iron to the pint, but possesses hardly any ferruginous taste. The establishment, when I visited it in September, had all the appearance of neglect. The pump for bringing up the water was out of order, the yard in which it was situated was over-grown with weeds, and the baths seemed uncared for, and were covered with a thick incrustation of oxide of iron. The resident physician, with whom I examined the spring, attached little therapeutic value to it, and I was told that this iron water was regarded by many as fictitious.

The mineral waters of Aix-la-Chapelle, though decidedly sulphurous, are seldom rejected by even the most fastidious stomach. Nay, strange as it seems that a fluid whose predominant flavour is that of rotten eggs, or the washings of a dirty gun-barrel, should ever be palatable, I observed that after the first day or two the majority of the water drinkers at the Elisenquelle appeared positively to enjoy their matutinal potations

The first effect of a glass of the thermal water is generally an agreeable sense of warmth in the stomach, which is succeeded, after an interval, by an increased appetite. The action of this Spa, whether it be used internally or externally, is that of a stimulant or excitant, operating principally on the kidneys and skin, and not, as is generally supposed, on the bowels. This determination to the surface and renal organs explains the efficacy of the water in many forms of cutaneous diseases, glandular enlargements, biliary obstructions, atonic dyspepsia, renal complaints, uterine derangements, impaired health from metallic poisoning, by mercury or lead; and in cases of lurking constitutional syphilis, in which it acts both as a test of the disease and as its remedy. The mineral waters of Aix-la-Chapelle are also prescribed in cases of chronic gout and rheumatism, rheumatic pericarditis and sciatica; and Dr. Velten informs me that he has seen benefit derived from their use in some forms of chronic bronchitis and catarrh.

Fully two-thirds of the invalid visitors to Aachen, however, as far as I could judge, suffer from cutaneous eruptions, or from chronic rheumatism. In these diseases the internal use of the water is combined with baths, which, indeed, are the chief part of the treatment here. The principal baths are the " Bains de la Rose," in the Comphausbad Strasse, which are supplied from the lower source, and are well-arranged and comfortable. The douch is peculiar; the attendant enters the water with the bather, and turns on the full force of the hot steam through the hose on the affected parts, kneading and rubbing them diligently with his disengaged hand at the same time. The temperature of these baths is 116 deg.

The "Bains Neuf," situated higher than those last described, are supplied from the upper source, and have a temperature of 119 deg. There are ninety-seven baths in this establishment, the atmosphere of which is intensely hot and humid. The Queen of Hungary's bath, close by, occupies the newest building, and is by far the best ventilated bath house.

I have already stated that these waters are powerful excitants, and need hardly add, that they are therefore countraindicated and dangerous in all cases in which a tendency to hæmorrhage, or apoplexy exists. They have been recommended in scrofula, but for the same reason I consider them unfitted for that disease. A resident physician, Dr. Wetzlar,* whose remarks are quoted with apparent approval by Dr. Edwin Lee,† holds the opinion that this thermal water is of great value in the treatment of progressive muscular atrophy, and creeping paralysis. But the pathological relations of these diseases should, however, lead us to

* Traite Practique Des Eaux Sulphureuses d'Aix-la-Chapelle—Bonn, 1856.
† The Baths of Rhenish Germany, p. 95. 3rd edition; London, 1861.

think that warm baths are not likely to be of the least use in such cases, and Dr. Velten, who has witnessed the ill consequence of this practice, confirmed my ideas, and told me of several cases of creeping paralysis, in which he believes great harm was occasioned by the use of these thermal baths. Gouty patients are frequently recommended to try this Spa, but unless the gout be perfectly chronic it will be to no purpose that they go through a course here, for experience shows that it is only in chronic and irregular gout that the mineral springs of Aix-la-Chapelle are serviceable.

The course of the *Aachen* baths and waters is usually six weeks, and few can continue it longer, as the effect is so debilitating that most patients can only use the douch twice a-week, and at the end of a course, even when cured of their original complaint, are so weakened as to require a short course of some chalybeate water, such as Schwallach or Spa.

Before leaving Aix-la-Chapelle, we visited the neighbouring springs of BORCETTE, or Burtscheid. This is a suburb of Aix, from which it is separated by the viaduct of the railway, and is a very quaint-looking old town, divided into an upper and lower quarter, consisting for the most part of dwellings of the poorer class, with the exception of eight or ten large bath-houses, a few hotels, and an ancient monastery of the ninth century, now the parish church, on the hill above the town. That we were in a place rich in thermal waters was evident the moment we entered Borcette, as through almost every street ran a river of hot mineral water, at which the inhabitants were washing their clothes. On every side we saw these thermal fountains steaming, and were at a loss where to begin our examination of them, till a French resident very kindly accompanied us to the several sources.

As in Aix-la-Chapelle the springs of Borcette are divided into the upper and lower sources. They are also divisible into sulphurous and non-sulphurous thermal waters. The superior springs are distinguished from the lower, and from those of Aix-la-Chapelle in not containing either sulphureted hydrogen gas or sulphate of soda, although resembling them in other respects. Of the sulphurous sources of Borcette, the most important is the Trinkquelle, at the foot of the hill, near the entrance of the town. The temperature of this spring is 138 deg., and it contains thirty grains of saline ingredients to the pint, twenty grains of which is chloride of sodium, two grains sulphate of soda, and six grains of carbonate of soda.

Amongst the non-sulphurous sources that are most generally used is the Kochbrunnen, or "boiling spring," in the centre of the town, rising from a rocky basin, eight feet wide. Its temperature is far from boiling however, being only 150 deg., while the Schwertbad source has a temperature of 170 deg., and its chief contents are chloride of sodium, carbonate and sulphate of soda.

## AIX-LA-CHAPELLE.

All the springs of Borcette are warmer than those of Aix, and instead of the white sulphurous precipitate of the latter they deposit a black, carbonaceous sediment. They are employed internally and externally, and are prescribed in cutaneous diseases—in dyspepsia and other gastric and intestinal complaints, in jaundice and in calculous affections. Moreover, the local doctors say they are applicable in some cases of scrofula, gout, and rheumatism. They are annually resorted to by close on twelve thousand visitors, mostly Germans and French. The chief reason for the preference which many of these give this place is, that it is a much cheaper residence than Aix-la-Chapelle.

## CHAPTER VIII.

#### FROM AIX-LA-CHAPELLE TO EMS.

Journey to Cologne A house on fire—Cologne—The Rhine—Oberlahnstein—The village wedding and our luggage—The valley of the Lahn—Ems: Description of the town—The thermal sources—Their composition and effects—Diseases for which applicable—Fatal influence of mineral waters in consumption—General remarks on this Spa.

HAVING missed the mid-day train from Aix to Cologne by a few minutes, we were obliged to wait six hours for the next departure, and so, this not being an express train, instead of arriving in time to see Cologne before dinner, as we had intended, we did not get to our hotel there till near midnight. Of the intervening country we saw hardly anything, as, until we reached Duren, the night was dark and cloudy. A little beyond that station, a lurid glare of red light suddenly disclosed a vast, solitary-looking plain extending for miles, and in a moment we passed by the origin of this light in a large country house on fire, close to the line. Momentary as was the glimpse afforded us, as the train shot by, it sufficed to show the upper part of the mansion in a sheet of livid flame; below were a few men running wildly about, apparently endeavouring to rescue something from the devouring element; a short distance off stood close together a little group of women and children, whose terror-stricken faces were distinctly visible in the red light, as they gazed on the destruction of what, till now, had evidently been their home. All this had passed before us in an instant, but not so readily did the impression produced by this scene wear away. Strong indeed must it have been, when it served to silence even the loud-voiced, noisy German ladies, who, till then, had maintained a shrill, incessant chatter, but now sat quietly for the rest of the journey. At all times, and under all circumstances, a great fire is justly regarded as a great calamity; still, after all, the destruction of a house by fire in a city, if unattended by any loss of life, or personal injury, is chiefly regretted as an inconvenience, a pecuniary loss, an untoward accident. But in the country it is something far more than this. Not only is it more dangerous for want of the aid which is at hand in cities, but, moreover, the feeling occasioned by the loss is very different. In town a house is ordinarily regarded only as a lodging, chosen either for fashion or convenience, while in the country there is generally some local attachment beyond this. The building may be old-fashioned, inconvenient, or even ugly in itself, yet these very faults are sanctified by their antiquity. In each incongruous addition we recognise some trace of the successive generations who have played their brief parts within the old walls, in every nook and cranny is some evidence of the tastes and fashions of our fore-fathers. When, therefore, an ancient mansion that has long withstood time and innovation, is suddenly swept away by fire, it is not merely so much stone and plaster destroyed, and furniture, plate, or books lost, as in town, but it is the annihilation of the "genus loci," the flight for ever of the Lares and Penates. And even where there are no such associations the destruction of a lonely country house by fire, in the stillness of the night, is infinitely more striking and appalling than a similar event amidst the noise and bustle of a large town.

I had just arrived at this point of meditation, when the gruff-voiced guard, who seemed by his manner to live in a perpetual passion, though in reality the most civil of Prussian conductors, demanded our tickets for the tenth time since we left Aix-la-Chapelle, and as he now carried them off we opined that we were near our destination. Shortly after, passing through a long tunnel, we emerged at the station at Cologne, and within ten minutes were taking our ease in our inn, the new "Hotel du Nord."

Of Cologne, its three kings, and cathedral, its waters, and odours, sweet and otherwise, I shall not attempt to speak, as these were not the waters which I came in quest of. Nor boots it to describe how we passed many days in slowly and quietly exploring the beauties of the Rhine, of which he who fancies he has "done the Rhine," by a day's excursion in a steamer from Bonn knows but little. As for those still more expeditious tourists who enjoy the Rhine from the windows of an express train on either of the railways that skirt the river, I can only say from experience that they cannot have the faintest idea of what the Rhine is like. For my own part, if travelling for pleasure, I would rather, albeit not much given to walking, follow the winding stream on foot from Rotterdam to Basle, than miss its beauties as those must do who content themselves with the flying glance the train affords.

Some days after our first arrival in Cologne, at eight o'clock of a bright, warm, autumn morning, we landed on the little pier of Oberlahnstein. The joy-bells from the village were ringing out a merry marriage peal, and I noticed a total absence of the stragglers who usually hang about such places. We were the only passengers that left the steamer, and no sooner had the porter landed our luggage than, leaving it on the ground without a word, he walked off. Nor could we induce him to carry it to the railway, as he was not a regular porter, all of whom, as he informed us, were at some wedding in the village. "Come now, good friend," we called after him, in the best German we could muster, "take up these two little portmanteaus to the station, and we'll pay you handsomely." At the word *trinkgeld* a grin passed over his broad face, and for a moment he hesitated between his German love of a spectacle and his cupidity, but he overcame the temptation, and gravely shaking his head, said "mein herr, it is impossible, for I am going to the wedding, too ;" and, lifting his greasy cap high into the air, with a most courteous salutation, wished us "guten morgen," and departed to the festival. Now the "little" portmanteaus we had invited him to shoulder were two of the largest solid leather articles of the kind that could be procured. Besides these, we were encumbered with hat boxes, carpet bags, railway wrappers, great coats, and all the rest of the usual incumbrances in the form of luggage considered essential for respectable travellers. How to transport all this to the railway station we knew not, nor even where the station might be.

For a moment, as we contemplated the pile of luggage, we almost gave way to despair, and wished our "impedimenta" at Jericho. This, however, did not mend matters, so, shouldering our portmanteaus, carpet bag in hand, we set out in search of the railway station, which was, fortunately, close at hand. Here we met the first persons we had seen since we landed. These were two individuals dressed in a handsome military uniform, who I took to be captains at the very least in the ducal army. One of these two wore an order on his breast, came up, asked where we were going, and before my wonder at the interest this gallant officer took in our movements was at an end, the other pasted labels on our luggage, the first proving the station-master, and the second the porter, and not a captain.

Twenty minutes by railway now brought us to Ems, and never did I travel in so luxurious a railway carriage. The highest praise I can give it is to say it was not unworthy of the dress of the guards. The line from Oberlahnstein to Ems turning inland from the Rhine, follows the winding valley of the Lahn through a most beautifully wooded country, formed by rounded hills and richly cultivated valleys for seven or eight miles, till it reaches Ems, where we arrived at noon.

Ems, which, with the exception of Carlsbad, is the most aristocratic watering-place in Europe, is situated on the right bank of the Lahn, about half-an-hour's journey by rail from the Rhine. It is divided into "Bad-Ems," the quarter around the springs, and "Dorf-Ems," the adjoining suburb. Bad-Ems consists of a row of houses built in crescent shape, between the river and the precipitous hills, which rise immediately behind it, and is connected with the railway-station by a handsome bridge, which has lately replaced a mediæval bridge of boats. Opposite the bridge is a kind of ravine in the mountain, and here a short street, the "Graben Strasse," has been built. A little beyond this, in the main street, is the large building containing the chief drinking spring; and the cursaal forming the extremity of the new quarter, from which an avenue of hotels and lodging-houses, fully a mile long, extends to Dorf-Ems, or the old village.

The mineral springs of Ems were known to the Romans, who have left unmistakeable traces of their presence round several of them.

A resident gentleman informed me that there are no less than twenty-five distinct medicinal springs here, and in some of the works on Ems there are detailed accounts of at least a dozen of these fountains, to each of which perfectly distinct, and, in some cases, opposite qualities are attributed. Only three, however, of these sources are used for drinking.

I shall first briefly describe the character of the Ems water generally, and then give the analysis of the most remarkable springs. Ems mineral water is strongly alkaline. It is clear and sparkling, and varies in temperature from 118 deg. to 83 deg. Its taste depends much on its warmth, the hotter waters having the Wiesbaden flavour of chicken broth, conjoined to a slightly chalybeate taste, while the cooler springs are more piquant from the larger amount of carbonic acid they contain, but are less agreeable.

COMPOSITION OF THE PRINCIPAL MINERAL SPRINGS OF EMS.

|  | Kesselbrunnen | Krähnchen | Furstenbrunnen |
|---|---|---|---|
| Temperature | 116° | 90° | 96° |
| Carbonic acid | 16·4480 | 26·8160 | 15·6760 |
| Carbonate of soda | 14·7418 | 12·6108 | 16·5526 |
| Carbonate of strontia | trace. | trace. | trace. |
| Carbonate of lime | 1·4474 | 1·4400 | 1·5263 |
| Carbonate of magnesia | 0·3200 | 0·4975 | 0·6206 |
| Carbonate of protoxide of iron | 0·0576 | 0·0096 | 0·0195 |
| Oxide of manganese | trace. | trace. | trace. |
| Sulphate of soda | 0·3538 | 0·3981 | 0·3678 |
| Choride of sodium | 7·0216 | 6·3349 | 5·8335 |
| Chloride of magnesia | 0·3318 | 0·3758 | 0·5248 |

Almost every Spa is recommended by the resident doctors as the best cure for every disease, and Ems is no exception to this rule. These springs are said to work wonders in cases of chronic bronchitis and pneumonia, the early stages of consumption, and "debility of the chest." As I do not exactly understand the meaning of the latter term, I shall express no opinion on it. But with regard to the statement which appears in several books, English as well as foreign, on this subject, that cases of consumption may be benefited by the mineral springs of Ems, I feel bound to express my strong doubts that any case of real tubercular disease of the lungs could have been cured by any mineral water whatever. On the contrary, I have no doubt that cases of phthisis have often been rendered more speedily fatal by the injudicious use of mineral waters.

The springs of Ems, being alkaline, saline, and alterative, are principally suited for cases requiring a mild, anti-acid aperient, increasing the secretions, and improving the appetite. They seem applicable to some cases of uterine ailments, and nervous diseases arising from these causes in females. They are, moreover, recommended in aphonia, hooping-cough, dyspepsia, and many other complaints.

## CHAPTER IX.

### SCHWALBACH AND SCHLANGENBAD.

Departure from Ems—Eltville—The Gasthoff—Moonlight scene on the Rhine—Drive through the Rheingau—Our companions—SCHWALBACH :—The town—Hotels—Resources for invalids—Account of each mineral source—The baths and the Empress of the French—Cudgeling versus cajoling—Analysis of the springs—Their physiological, pathological, and curative effects—Diseases in which they should be used—Account of SCHLANGENBAD—The waters and baths and their uses.

OUR landlord in Ems had forewarned us that we should not be able to reach Schwalbach on the afternoon we left his hotel, but, disregarding his assertion, we started, and got to Eltville late in the evening, only to find the prediction verified. There was a coach to Schwalbach, but as there was no vacant seat, it was unavailable for us. In vain we tried for some other conveyance. There was none to be had. No other train was expected, the railway officials went home, the station was shut up, and we were left on the platform. So, again shouldering our luggage, we set off, Syntax like, in search of an inn. Hardly had we left the station, when the Railway Hotel presented itself, and, as there was no other building in sight, into it we entered, and demanded " zwei zimmers," but as there was but one spare bed-room we had to content ourselves with an uncarpeted, whitewashed apartment, ten feet long by six feet wide, luxuriously furnished with two truckle-beds, a couple of deal chairs, and washing stands, containing each a small croft of water, and a diminutive pie-dish for basin. There seemed no choice, so we were obliged to follow Lord Russell's counsel, "rest and be thankful;" and accordingly we ordered dinner and a bottle of Gräfenberg, for we were now in the capital of the Rheingau, the "Bacchanalian Paradise of the Rhine" as the poet calls it. Meantime, while our repast was preparing, we strolled through the town, which is some distance from the railway station, on the Rhine, and in the course of our perigrination saw one or two tolerable-looking hotels, that contrasted very strongly with our *Posada*, but it was then too late to change. The little town has an appearance of considerable antiquity, and is composed of a few dark, narrow, winding lanes. Our appetites now began to warn us that we were wandering too far from our inn, to which we accordingly returned, and were regaled in the following manner: first was introduced a tureen of Brobdignagian dimensions, and sounding its depths with a leaden ladle we discovered its contents to be a thick, yellowish soup, made of pumpkins. To this succeeded pounded "schwein-fleisch," compressed in a mould, in the form of cutlets; the inevitable "braten," or roast, came next, and this was a small, shapeless joint, which might have been anything, but which was, I believe, meant for "kalbfleisch," or veal, garnished with a conserve of prunes, and the repast terminated by a "kartoffel salat," or potatoe salad. As for the "Gräfenberg," we concluded that it must have been drawn from the same cask as the vinegar with which the salad had been dressed. But, as the Spaniards well say, *Al hambre no hay mal pan*, " to the hungry man there is no bad bread," and such was our experience even on this occasion.

After dinner we again walked down to the river side, where, seating ourselves at the end of the little pier, we smoked a peaceful calumet. The moon had attained her full height, unobscured by a single cloud, and shed a soft light that rendered every object as visible, but far more beautiful than in the day. Before us lay stretched the broad Rhine, here expanding into a lake; not a ruffle stirred its quiet waters, and not a sound broke the stillness of the night. Behind was Eltville, the grey towers of whose ruined walls assumed a more imposing appearance in the moonlight, and contrasted strongly with its low dark streets. As we contemplated this fair scene, Geibler's legend of " Charlemagne Blessing the Rhineland," so admirably rendered by that ill-appreciated, and most unfortunate of modern poets—Clarence Mangan, came before us with new force and feeling :

"Beauteous is it in the summer night, and calm along the Rhine,
And like molten silver shines the light that sleeps on wave and vine.
As a bridge across the river lie the moonbeams all the time,
They shine from Langawinkel unto ancient Ingelheim ;
And along this Bridge of Moonbeams is the Monarch seen to go,
And from thence he pours his blessing on the royal flood below."

For fully an hour had we sat here undisturbed, when

at length we heard a distant bell, towards Biberich, and shortly after heard the paddles of the approaching steamer. A few boatmen now came out of the old gate of the town and joined us on the pier; but as yet the steamer was invisible, though heard for miles in the still night, till at last, through the mist now rising from the river, we saw its black hull slowly nearing us. It stopped opposite to us, a boat put off, and, after a hard pull against the current, one of the passengers was landed, and we were again left to our solitude.

The church clock behind us now tolled out eleven, and considering that though moonlight musing by the Rhine was very agreeable, still, as rheumatism the next morning might not be equally so, I suggested to my companion that it would be wiser to move away. Not, however, as yet feeling disposed to seek the shelter of our inn, we struck out into the country, and walked a couple of miles through the orchards before returning to the " gasthof."

Next morning we rose betimes, to secure seats by the first coach to Schwalbach, and, but for this precaution, should have fared as we had done the preceding night, as the coach was already nearly filled. The road which lay through Schlangenbad, was long and tedious, the day was sultry and dusty, the coach was full, and the windows were all shut, so at the first stoppage we made, I mounted on the roof, despite the remonstrance of the conductor, who insisted there was no seat, while I proved the contrary by perching myself on a portmanteau; but soon found that I had little reason to rejoice in my elevation over my fellow travellers. For the suns rays became so scorching that even our driver had hardly sufficient energy left to flog the jaded horses, the conductor had deserted the coach and was playing the gay Lotharia with a country girl who was driving a team of cows, while the portmanteau upon which I sat having slipped away from its moorings, had to be held on the roof, with no small difficulty at every jolt. Anxiously did I look up the long hill before us, until at length we reached the plateau at the summit; but no trace of Schwalbach was yet visible. Suddenly, however, an unexpected lurch nearly threw me from my precarious seat, as, for the first time, the horses broke into a brisk trot, and turning down a steep hill, we found ourselves in the town, which, lying in a narrow valley, is thus hidden from view. Schwalbach wore its gayest aspect. We passed under a triumphal arch, every house was decorated with flags, and bands of music were playing on all sides. What the cause of rejoicing might be we knew not. Could that surly German inside, who smoked such execrable tobacco, and used a fore-finger long ignorant of soap and water, but adorned with a cameo ring as large as a crown piece, instead of a pipe stopper, be the Grand Duke of Nassau, in disguise? Or was his vis-á-vis, a sallow youth, with an incipient moustache, a " wide-awake" hat and a Dundreary tie, whose stock of language seemed limited to two French oaths, both ill pronounced, some noble lord, whose advent was anxiously expected? Thus I puzzled myself, till a banner bearing " Vive l'Imperatrice " caught my eye, and brought to mind that the French Empress had been that day expected in Schwalbach.

Alighting at the Nassauer-Hof, we proceeded to call on the principal physician of the town, Dr. Genth, to whom I had a letter of introduction, and finding the Doctor at home, he very kindly gave up several hours to showing us over the various objects of interest in the town and its environs.

Before accompanying Dr. Genth through the springs and baths, a brief coup d'œil over the town will not be misplaced.

*Langen-Schwalbach*, as it is called, to distinguish it from a neighbouring village of the same name, also containing a mineral water, is said to have been known as a favourite watering place to the Roman legions stationed in Germany. Be this as it may, however, it is certain that for ages it has been largely resorted to by invalids from the neighbouring countries, but it was hardly known in England until Sir Francis Head published his " Bubbles from the Brunnen," in 1830. And so great has been the effect of his enthusiastic eulogy, that during the last season, as Dr. Genth informed me, there were upwards of 600 English valetudinarians in Schwalbach.

The town itself, contains between two and three thousand resident inhabitants, and is situated in a little valley sunk between the lofty hills which form the table land of the Taunus range. Schwalbach lies some 900 feet above the level of the sea, and is about ten miles distant from Eltville, about twelve from Wiesbaden, and double that distance from Ems. The name " Langen " is well applied, for the entire place consists of a long straggling street, expanding near the centre into a kind of open crescent, with two short diverging streets, the whole being built somewhat in the shape of a capital Y.

A little in front of the main street is a small garden or park, connected with the bath house, which is the principal building in Schwalbach, and is a large, commodious, but not very ornamental structure. In front of this is a colonnade, containing shops and stalls, and on the upper floor are the reading rooms, which in bad weather are the only resource open to visitors. The higher part of the street consists of large detached houses, nearly all of which are either hotels or lodging houses. The preponderance of French and English visitors may be inferred from the advertisements on the houses, which are generally in either of those languages, and very rarely in German. In the lower part of the street the houses are of an inferior kind, and a few small shops may be found.

There are plenty of hotels, and perhaps one of the best, though not the cheapest, is the

"Nassauer Hof," the cuisine of which is much above the average of German hotels. Most of the invalid visitors, however, go into lodging-houses, of which there are some forty or fifty in the upper part of the town. And, considering the fashionable character of this watering-place, the prices, both of the apartments and the living, are exceedingly moderate.

In the way of amusements, Schwalbach has not much to boast of; excellent bands play at stated hours in the Allée, there is a pretty park and miniature lake, reading and billiard rooms in the village, and charming walks and drives in the vicinity, *et voila tout.*

The principal mineral springs of Schwalbach are the "Weinbrunnen," the "Stahlbrunnen," and the "Paulinenbrunnen."

The Weinbrunnen is placed on a little hillock between the Bath-house and an hotel, occupied at the time of our visit by the Empress of the French; the source is contained in a sunken area, a few feet under the surface; and is enclosed by a handsome circular balustrade. The water, which is clear and sparkling, appeared in a violent state of ebullition from the large amount of carbonic acid gas it contains. Its taste is agreeable, piquant, and ferruginous. This source is the oldest of the Schwalbach springs used medicinally, and was first brought into vogue by Dr. Theodore, of Worms, surnamed Tabernæmontanus, from his place of birth, who in 1581 extolled its use as a remedy for the unpleasant feelings experienced the morning after an extra cup of wine, and hence is derived its name.

The Weinbrunnen is considerably stronger in iron than either the Stahl or Paulinenbrunnen, but containing less carbonic acid gas, and more intimately combined with it, this water is less exciting, and agrees better with some constitutions than the other sources.

We now crossed to the south side of the town, to the Stahlbrunnen, which is situated at the extremity of a shady promenade—the Allée. The taste of this spring is very similar to that of the Weinbrunnen, but is still more gaseous, and as the gas is less intimately united to the iron, the ferruginous flavour is rather more perceptible. I moreover noticed a slight taste of sulphureted hydrogen. The temperature of the water was 50 deg., or one degree higher than the last spring.

From the Stahlbrunnen, a very pretty walk by the lake through the Allée, over which the foliage formed a complete arch the entire way, brought us to the *Paulinenbrunnen*, of which every reader of the " Bubbles from the Brunnen,"—and who has not read them?—will recollect Sir Francis Head's glowing description; but, alas, were Sir Francis now to revisit his favourite spring, he would hardly recognise in it his own picture. For ever since Dr. Fenner's death, some ten years ago, the *Paulinen* has been declining in favour, till it has now reached the acmé of its neglect. At the time of our visit the railing that surrounds the court in which it rises was unpainted and covered with rust, the steps leading down to it were broken, the flags were forced up by the thick vegetation growing between them, there was no one to give out the water, and the only seat in the place was a broken deal table, which lay upset in the middle of the court. During the hour we sat there musing —Marius-like, on the fallen fortunes of the Paulinen— not a single person approached this source, though the town was crowded with visitors, and it was the hour for the afternoon potations at the springs. The Paulinen rises from a shallow stone basin, the bottom of which was incrusted by a deposit nearly two inches thick of carbonate of iron. The water is not by any means as clear or sparkling as either of the sources I have already described, though in reality it contains much more gas; but this seems to be less diffused through the water, and comes up to the surface in large bubbles. Its temperature was two degrees lower than the Stahlbrunnen, or only 48 deg. Being the mildest of the springs, it therefore agrees in some cases in which the stronger waters of the Wein or Stahlbrunnen would not be tolerated. Its taste, however, is not agreeable, being ferruginous and slightly saline. In the same enclosure is the " Rosenbrunnen," but as this is only used for bathing purposes, I need not describe it more minutely. Besides these, there are six other mineral springs in Schwalbach, but they are either not used or else are identical in composition with those I have just spoken of.

We now returned to the Bath-house, where Dr. Genth had an appointment with an exalted personage. At the door we found a considerable crowd assembled, and none were allowed to enter, as the Empress Eugenie was then in the bath; but the doctor being recognised, was at once admitted, and I followed closely in his wake to the ante-room, where a number of ladies were awaiting Her Majesty. In a few minutes the door was thrown open, and the *Comtesse de Pierrefonds* swept out, flourishing a small cane in her hand, and such is the peculiar imitative proclivity of the female mind, that from that time every lady we met brandished a ratan before our eyes. Nature has indeed been sufficiently lavish of her offensive and defensive armoury to the fair sex, and art would find it difficult to invent weapons more killing than their eyes, or sharper than their tongues. Of old they were wont to commit sufficient havoc with these, leaving the cudgel to their brothers and fathers; but what chance would the ruder sex have were the ladies, not satisfied with cajoling, now to resort to cudgeling them?

Nothing could be more richly decorated than the bathing rooms that the Empress had just quitted, which, up to a few days previously, had been reserved for the Empress of Russia. But the ordinary bath rooms, though lofty and airy, are in no wise handsome. The head of the establishment conducted me through it, and explained the various arrangements. There are three stories in the house, the baths of which are supplied from different sources. Those on the ground floor are filled from the Rosenbrunnen, the second floor from the Stahlbrunnen, and those on the upper story from the Paulinenbrunnen.

The contrivance by which the mineral water is brought to a proper temperature for the bath, without undergoing any change, or having its carbonic acid gas driven off, is very ingenious. All the baths are incased within metal boxes, so constructed as to leave a space of about four inches round the bath; into this space steam is forced by an engine below stairs, and thus applied is capable of raising the temperature of a large bath full of water from 50 to 90 degs. in less than ten minutes.

Some idea of the quantity of mineral water consumed in this establishment, in which there are about sixty bathing cabinets, may be formed from the fact, that during the last two seasons, the number of baths amounted to three hundred and fifty a day, and sometimes, including the private baths, reached nearly four hundred, which at the rate of fourteen cubic feet per bath, would require a daily supply of five thousand six hundred cubic feet of mineral water. Now a recent writer, Mr. Pole, has demonstrated that the three springs which supply the baths can together only furnish four thousand eight hundred cubic feet of water per diem, and whence the remaining quantity of mineral water is obtained, I was not able to ascertain.

In considering the physiological effects of the Schwalbach mineral springs, and how it happens that they are often far more beneficial than pharmaceutical preparations of iron of the same strength would be, it should be borne in mind that in these springs the iron is contained in perhaps its most soluble and easily digested form, namely that of the carbonate of the protoxide, and that moreover it is combined with a large quantity of a very important medicinal agent, carbonic acid gas. By the first of these principles the character of the blood is altered, its hæmatin and red globules are increased, its crasis becomes greater, and it is better fitted for performing its vital and stimulating functions on all the structures and organs of the body. The carbonic acid produces not less important effects, acting as a powerful excitant of the functions of the nervous system. Dr. Genth, who has devoted considerable research to this subject, is of opinion that the carbonic acid gas has moreover the effect of rendering the blood corpuscles less soluble than they were previously, and thus it too indirectly contributes to the increased crasis of the vital fluid which generally attends the administration of the Schwalbach waters.

In the following table I have nearly followed Dr. Sigismund Sutro's analysis of the

COMPOSITION OF THE SCHWALBACH MINERAL SOURCES.

| Contents. | Wein-brunnen. | Stahl-brunnen. | Paulinen-brunnen. | Rosen-brunnen. |
|---|---|---|---|---|
| Carbonate of protoxide of iron | 0·83 | 0·75 | 0·65 | 0·91 |
| Carbonate of lime | 2·11 | 1·45 | 2·95 | 2·95 |
| ,, ,, magnesia | 3·12 | 0·88 | 2·75 | 0·98 |
| ,, ,, soda | 0·17 | 0·25 | 0·45 | 0.35 |
| Chloride of sodium | 0·18 | 0·34 | 0·03 | 0·32 |
| Sulphate of soda | 0·16 | 0·21 | 0·02 | ... |
| Total solid contents | 6·59 | 3·83 | 6·86 | 5·51 |
| Carbonic acid gas | 26 cub. in. | 28 | 39½ | 26 |

According to the last analysis by Fresenius, the red ochreous deposit found in all the springs, when dried at the temperature of 212 deg. contains:—

| | |
|---|---|
| Oxide of iron | 60.06 per cent. |
| ,, ,, manganese | 0·39 ,, |
| Phosphates | 1·04 ,, |
| Arsenites | 0·0137 ,, |
| Silicates and insoluble residue | 23·10 ,, |
| Water | 13·20 ,, |
| Lime, magnesia, byrites, strontium, oxide of copper, oxide of lead, carbonic acid, and organic matter | 2·2963 ,, |
| Total | 100.000* |

In nearly all cases in which a ferruginous water is required, the springs of Schwalbach have been recommended, and there is now no continental chalybeate Spa so much resorted to by English invalids. And when we consider the vast number and variety of ailments in which an unhealthy and anæmic condition of the blood is a leading feature, we must admit that a

\* "Les Sources Ferrugineuses de Schwalbach, par Dr. Genth." 2nd Ed. p. 44. Schwalbach, 1860.

mineral spring which holds out a promise of cure in such cases deserves considerable attention.

Three fourths of the invalid visitors to Schwalbach suffer from some form of anæmia, amongst whom may be found a great many ladies labouring under chlorosis, or whose ailment is weakness and want of tone, following perhaps a London season, and for such patients there can be no mineral water of more general utility. For in these cases the quiet of Schwalbach is perhaps not less useful than its chalybeate springs. In the debility following convalescence from severe and exhausting maladies, and in the weakness resulting from long protracted nursing, the Paulinen or Weinbrunnen, sometimes even diluted, are often serviceable.

Dr. Genth recommends a trial of this chalybeate in some cases of infantile wasting, and in bronchitis; but the number of such cases which would be served by a journey to the Brunnens, is, I think, extremely limited. Nearly all authorities on the subject speak of the powerful action of this Spa in certain forms of functional derangement of the female system, and my own experience in some instances leads me to share this opinion.

In chronic indigestion or dyspepsia, consequent loss of flesh, and constipation, depending on a relaxed bloodless and torpid condition of the mucous membrane of the stomach and alimentary canal,—so often the cause of these common maladies, which the drastic remedies, the pills and draughts, that many people take habitually, can only aggravate—a visit to Schwalbach may be prescribed. These ailments, in most cases, yielding to mildly stimulating chalybeate, and somewhat saline mineral waters such as these.

The mode of drinking the mineral waters is much the same here as at Ems or Spa; it should be taken fasting in doses of from half a glass to two tumblers at a time, thrice a day, and should be followed by a brisk walk. Formerly much larger doses were prescribed, for instance, as Dr. Genth informed me, from sixty to seventy ounces daily—such quantities are now happily never administered; though I have seen a gentleman consuming as much as four large glasses full of the water in succession.

Like all other remedies of any real efficacy, the waters of Schwalbach, when not required, cannot be used with impunity; and, in certain conditions of the body, might even, give rise to fatal consequences. They should never be employed by plethoric patients of sanguineous temperament, in whom inflammation is easily lighted up. Nor should they be drank by invalids, who have any tendency to hæmorrhage or organic disease, whether of the lungs, heart, or kidneys. In short this water is not one of those remedies, which, if it does no great good, can at least do no harm; on the contrary, it often, when improperly used, leads to very injurious results, and therefore should never be resorted to unless with the sanction of the patient's medical adviser.

From Schwalbach to Schlangenbad the transition is easy and natural, for only five short miles separate them, and yet how different are these Spas. Of the beauties of Schlangenbad, so many charming pictures have been drawn since the " Bubbles " were blown, that I fear to venture on any but the most prosaic description. So smitten was Sir Francis Head with its situation, that he says, " I never remember to have existed in a place which possessed such fascinating beauties."

Schlangenbad lies midway between Eltville and the Spa I have just described, in a valley almost hidden amongst thickly-wooded hills. It can hardly be called even a village, for it consists merely of a few bath houses, hotels, and some twenty or thirty large barrack-like lodging houses, irregularly scattered through the little valley. Being situated on the south-western slope of the Taunus range, and well protected from harsh winds by the hills, it enjoys a much milder and more genial climate than Schwalbach. There is in Schlangenbad a total absence of the life and gaiety that characterizes almost all the Spas of Rhenish Germany, in place of which there is here a kind of sleepy atmosphere, which, irksome, and even injurious as it would prove to many, must, I am sure, be most useful in some cases of nervous irritability and excitement, resulting from extreme mental tension, and over attention to any absorbing pursuit in busy civic life.

Apart from this placid tranquillity, this " dolce far niente " existence, there is little or no attraction in Schlangenbad for any but real valetudinarians. For beautiful as is the scenery and interesting as are the excursions in its vicinity, they are almost equally accessible from Schwalbach or Wiesbaden. The only attempt at dissipation is the excellent band which twice a day performs on the little terrace adjoining the bath house. Our dinner at the Nassauer-Hof was served in the veranda overlooking this terrace, and good as we found the golden-coloured, fragrant Raunthaler, not a little of its zest was due to the music floating up from below, intermingled with the splashing fountain and the distant hum of voices on the promenade, by which it was accompanied.

On this promenade I once witnessed an instance of German good taste. As usual the band was playing after dinner, and all the visitors were assembled, the ladies plying their never-ending embroidery, the children playing around them, and the men all sending up volumes of dense smoke, like so many ambulant factory chimneys. A carriage drove up, and the Duke of Nassau, the Empress of the French, Prince Metternich, and one or two ladies entered. Every one took off their hats as the party passed, but beyond this no notice was taken of them, no one followed them, nor was there any rush to stare at them. Now certainly this must have been more agreeable to these personages than to be stared at, cheered, and followed, as would probably have been the case were a similar party to promenade through an English watering-place.

The waters of Schlangenbad belong to the same class of mineral springs as those of Pfeffers and Wildbad, being, however, stronger than either of these Spas. They are mildly alkaline and thermal, and are chiefly employed for bathing purposes. There are eight distinct springs, which vary in temperature from seventy-seven degrees to ninety degrees. They rise at the foot of the adjacent mountain whence they are conducted to the bath houses. They all contain about eight grains of solid ingredients, with two cubic inches of carbonic acid gas, and the same amount of nitrogen in the pint. Of the solid constituents about one-half consists of carbonate of soda, together with two grains of common salt, and one grain each of carbonate of lime and carbonate of magnesia. The special action of this Spa seems to be on the skin, which it is said to render soft and white, and therefore I need hardly add is largely patronized by the fair sex. This property however, can only result from the alkaline constituents of the water combining with the fatty acids of the skin, with which they form a soluble soap, and therefore the effect is merely temporary, and is more likely, ultimately to injure than to serve the skin.

The power of diminishing nervous irritability has been ascribed to these waters by many writers, and they are, moreover, largely employed in the treatment of chronic rheumatism, as well as neuralgia and other nervous affections. In cutaneous diseases, too, such as lichen, and prurigo, when a remedy is required to allay excessive irritation of the surface, a course of the Schlangenbad baths is often prescribed with advantage. Hufland and other German writers recommend these baths in the articular rigidity of advancing years, as the veritable "Fountain of Youth" of the fairy tales. As a specimen of the rhapsodies, which German Spa physicians sometimes indulge in, I shall conclude this chapter with the following *morceau* from the late Dr. Fenner, of Schwalbach, who, in describing the effects of this bath, thus falls into an ecstacy of praise—" Vous sortez des eaux de Schlangenbad rejeuni comme un Phœnix—la jeunesse y devient plus belle, plus brillante, et l'age y trouve une nouvelle vigeur."

## CHAPTER X.

### WIESBADEN.

*Arrival in Wiesbaden—A cheerful prospect—A cab stand versus an hotel—"Unprotected females"—The author's captivity and deliverance—Description of WIESBADEN—The Kochbrunnen—Analysis of the springs—Mode of using the water—Its effects—Diseases in which it is prescribed—The hot baths and their sanative influence—Cases in which this Spa should be interdicted.*

After a long journey the prospect of passing the night in a damp, musty hackney coach on a stand, instead of a comfortable bed in an hotel, is not particularly pleasant. But on the night of our arrival in Wiesbaden such seemed to be our destined fate. We had tried almost every hotel in Wiesbaden, and all were full; and having exhausted our list, had made up our minds to remain where we were till morning. "Where shall I drive to next?" asked the driver. "Nowhere," I replied, "we'll hire you by the hour, so stay where you are; *gute nacht*, good night." "But I want to go home to bed," expostulated Jehu. "So do we," said I; and without another word our driver sprang on the box, and, lashing his horses, drove at a gallop through the town, till, pulling up before a large building, he entered, and presently emerging, he informed us that they had rooms in the *Hotel de France*, which we gladly secured. Hardly, however, had we installed ourselves when the landlord made an appeal for one of our rooms for two ladies, "compatriotes," as he informed us, who had just arrived. Of course we consented to give it up, but reminded our host of the slight objection that the rooms communicated by folding doors, and that to enter one chamber it was necessary to pass through the other. "*Cela ne fait rien,*" he replied, "you gentlemen can remain in the inner room till the ladies are gone out in the morning." So the matter was arranged, and while we were collecting our scattered effects, two severe and decidedly serious-looking elderly ladies entered, who, with the smallest possible acknowledgment, took possession of our room. Hardly had we retreated to the second chamber than we heard ourselves securely locked in, and, as there was no other means of exit but the door which was thus fastened, excepting an inaccessible skylight, we had the satisfaction of reflecting that should the house take fire during the night, we, at any rate, would die martyrs to our politeness: nor was it till near noon next day that we were released from our durance. It was then, of course, too late to present my letters of introduction as the early morning is the usual time for calling on the German physicians. This loss was, however, compensated for by the attention of our fair neighbours, who, when they had secured other rooms, sent us a souvenir of their gratitude in the form of a packet of light and interesting tracts, including, " Crumbs of Comfort," " The Terrified Trumpeter," " A Timely Warning," and several other equally valuable pamphlets.

In the first place I proceeded to pay a few visits to some friends, who were staying here for the season, and under their guidance, before examining the mineral waters, devoted a day to making myself familiar with the topography of the town, and the following were the notes which I jotted down on the spot :

Wiesbaden enjoys a situation that renders it one of the most picturesque of the German Spas, and which should also bless it with a climate superior to many of them. Situated in an opening valley, extending from the southern slope of the Taunus mountains, at the base of which the town is built, to the Rhine, and thus completely protected from the north and east winds, Wiesbaden lies some twenty-five miles to the south-east of Frankfort, and ten miles south of Mayence. The first view of the town is certainly prepossessing : it occupies a considerable expanse of ground; some of the streets are wide and clean ; the buildings are large, bright, and new looking ; and the square in front of the cursaal, and that edifice itself, are really handsome. But in no sense of the word is Wiesbaden, although the capital of the Duchy and the seat of Government, a city ; it is merely a large watering place, which is best described by Sir Francis Head's designation of a " city of lodging houses ;" and considering that the resident population is barely fifteen thousand, and that the number of visitors during the summer often exceeds thirty thousand, the great majority of whom reside in lodgings, there must be a great deal of truth in the title.

Although in reality the most ancient of the German watering places, every building in Wiesbaden seems fresh from the workman's hands. The principal part of the town consists of hotels, baths, and lodging houses, together with a few public buildings; namely a gambling house, post-office, and theatre, intended mainly for foreigners.

The hot springs of Wiesbaden were resorted to by invalids at a very early date. Pliny thus describes them, " Sunt et Mattiaci in Germania fontes calidi trans-Rhenum, quorum haustus triduo fervet; circa marginem pumicem faciunt aquæ."* It seems moreover well proved that these hot springs were used by the Romans long before Pliny's time. The remains of numerous " Balnearia" have been discovered in the vicinity of the mineral sources, and that they were not used merely for cleanliness or luxury, but as a remedy, is attested by the votive tablets to Apollo found in these " Thermæ," some of which are still used as baths.

There are no less than twenty-two thermal springs in Wiesbaden. Some writers describe each of these separately, and assign different properties to each, but

* Pliny Hist. Nat: Lib. xxxi., c. 17.

as all issue within an area of three thousand yards, and vary chiefly in temperature, in all probability they all originate from the same source and, at any rate, differ so little, that details of each would be quite needless, and therefore in the following remarks I have confined myself to the Kochbrunnen, which I regard as a type of all the other springs.

The Kochbrunnen rises nearly in the centre of the town, in a large shallow basin, walled in, and only accessible in front by a gate—at which sits a very ugly specimen of German womankind, dispensing the steaming fluid. This is conveyed by metal pipes across the basin, from its origin on the opposite side to two shell-shaped vases, from which the glasses are filled. The appearance of the water is far from inviting, having a turbid, yellowish colour, with a scum floating on the surface. The taste, however, is by no means unpleasant, and Sir Francis Head's comparison of it to "weak chicken broth" has been copied by every succeeding author as conveying an exact idea of its flavour. Sir Francis Head was, probably, however, not aware how fully the Kochbrunnen deserves the appellation of "broth," for it has been since proved that it contains large quantities of living animalcula, which perhaps may impart the peculiar taste all writers on Wiesbaden have descanted on.

The temperature of this source is differently stated by authors: Dr. Granville says it is one hundred and fifty-five degrees; Dr. James Johnson, one hundred

WIESBADEN. 151

and sixty degrees; Dr. Wilson, one hundred and fifty-five degrees; and I found it one hundred and fifty degrees. Probably the explanation of this discrepancy may be, that the temperature of the water is not always the same, and not that there is any inaccuracy in the observations.

A pound of Kochbrunnen water contains sixty-three grains of solid matter, and of this no less than fifty-two grains are common salt, or chloride of sodium. The other principal salts in it are chloride of calcium, or lime, of which there is nearly four grains, carbonate of lime nearly three and a half grains, chloride of magnesium, one and a half grains, and rather more than a grain of chloride of potassium. Besides these, it contains sixteen other ingredients, but in such small proportions as to be of no practical importance. There are only six and a half cubic inches of carbonic acid gas in a pound of this water.

The thermal waters of Wiesbaden are principally employed for bathing; but as the Kochbrunnen is also extensively used internally, a few observations on its effects when employed in this way are necessary. This source is daily frequented by a considerable number of invalids, most of whom seem to suffer from gout, or dyspepsia, or hepatic disease. Under the long iron gallery that shelters the water drinkers from the rain, though not from the wind, plethoric aldermen from Cheapside, oppressed by calipash and calipee, yellow Indian officials, with congested livers, brought on by tropical heat, curry powder, and improper diet; and gouty-looking votaries of good living and indolence, saunter up and down between the prescribed doses of the water.

Like all other mineral waters, this should be taken fasting, before breakfast. The dose varies from eight to thirty ounces of the Kochbrunnen, which is sipped slowly while the patients promenade about the spring, and the exercise, I have little doubt, does almost as much good as the water.

The action of Wiesbaden mineral water, taken in doses of from ten to twenty ounces, is first shown by an increase of all the secretions of the drinker, particularly those of the mucous membranes, a mildly purgative and diuretic effect being generally experienced. The circulation is always more or less excited by it, the biliary secretion is augmented, the action of the absorbents is quickened, and the appetite is sharpened, though this should not be indulged. If the water be now persisted in for some weeks, the bulk of the body diminishes visibly, the expanded abdomen subsides, obesity disappears, and the outlines of muscles, previously concealed by superabundant fat, are thus brought into view.

After some time the blood generally becomes thinner and less rich in fibrinous compounds, respiration is now freer, the skin becomes clear and healthy looking, and the valetudinarian experiences a general feeling of *bien etre*.

Such are the effects of the Kochbrunnen when it agrees with the patient. Unfortunately, however, it does not answer every case, though some of its panegyrists seem to think it does. When the water does not produce its purgative or diuretic effects, and especially in patients suffering from any organic disease of the heart or lungs, or those of an irritable habit of body, active inflammation may be readily induced by this spring, or formidable hæmorrhage brought on.

When a course of the Kochbrunnen is prolonged beyond six or eight weeks, symptoms of what Dr. Braun calls "saturation" set in. "The drinker experiences an aversion to the thermal water, which up to this period he has taken with pleasure. . . . . The secretions become irregular, the tongue covered with a whitish substance, the taste becomes languid, and the appetite feeble; the thirst becomes more intense, and there ensues a general feeling of relaxation and fatigue."* The local physicians say that if these symptoms be neglected, what they term "super-saturation" will be produced, and may lead to very disastrous results. As I have already treated of this subject very fully, I need not enter on it here.

The diseases in which the internal use of the Wiesbaden waters are recommended are gout, dyspepsia, and plethora. In the irregular and atonic forms of gout, and especially in that constitutional condition

* *The Hot Springs of Wiesbaden.* By Charles Braun, M.D., p. 53 Wiesbaden. 1856.

which I have described under the name of gouty diathesis, the thermal waters of Wiesbaden often produce great benefit. Repeated experiments have shown that the amount of urea and uric acid eliminated by the kidneys is greatly increased under their use. This fact in some measure explains the occasional curative action of the water in cases of gout, in which disease the blood is loaded with uric acid and urates. The general health is also improved by the alterative action of the Spa; the secretions being all increased, the abnormal matters, long retained in the system, are set free; and the appetite and strength are repaired. Thus the protean forms of gout, simulating every disorder, and often affecting the mind through the body, sometimes changing a man perhaps naturally of an amiable disposition and cheerful temperament, into a peevish hypochondriac, are occasionally permanently cured, or more frequently are mitigated, by a course of Wiesbaden water. In some cases of lurking gout, this water brings on a fit of regular podagra, which is generally attended with complete relief of the other symptoms.

In cases of dyspepsia connected with an acid condition of the stomach, loss of appetite, slow and imperfect digestion, and constipation dependent on atony, or intestinal weakness, these waters are frequently administered with the best possible results. In such cases they should be taken in full doses, and as hot as is agreeable to the patient.

I have already mentioned that the thermal springs

of Wiesbaden are principally used for bathing purposes, and the number of baths in the town is something extraordinary, amounting to nearly nine hundred. Almost every hotel has its thermal department, supplied from the twenty-five sources I have spoken of. Most of the baths that I visited were well arranged and clean; each bath is provided with two conduits, one of which brings the thermal water direct from its source, and the other conveys the same water previously cooled. On entering one of these cabinets, the visitor cannot fail to be struck with the thick incrustation of calcareous matter round the sides of the bath; and the turbid and peculiar appearance of the water, which is covered with a whitish film.

The bath should either be taken in the morning fasting, or in the evening, five or six hours after an early dinner. The patient at first should not remain more that ten minutes in the water, but may gradually, if so advised, increase this period to an hour.

These baths are principally employed in cases of rheumatism. According to Dr. Braun, nearly one half of the entire number of valetudinarians who resort to Wiesbaden, suffer from rheumatism, or from that combination of gout and rheumatism, permanently affecting the joints, which is known as chronic rheumatic arthritis. Of one hundred and twenty-nine cases of chronic rheumatism treated by Dr. Haas in the hospital of Wiesbaden by Spa water, above thirty were completely cured, and seventy-nine were improved. The German physicians also recommend this remedy in rheumatic paralysis, where the power of voluntary motion is lost, and parts are forced into unnatural constrained positions, by long continued arthritic inflammation.

## CHAPTER XI.

#### WEILBACH AND SODEN.

The Journey—Orchards of Nassau—Traits of the Peasantry—WEILBACH—A High-Dutch Amazon—The Mineral Springs and Baths—Monotony of Spa-life here—Cases in which the Weilbach Water is employed—Route to Soden—Harvest Scene—A Contrast—Account of SODEN—Great Variety of Mineral Springs—Their Composition and Uses.

FROM Wiesbaden an hour's drive by train, through a richly-cultivated vine-country, and affording *en passant* a glimpse of the spires of Mayence, brought us to the little village of Flörsheim, where we alighted at nine a.m. of a bright sunny morning. Hence, half-an-hour's walk through a continuous orchard conducted us to the baths of WEILBACH. Short as this walk was, it was yet sufficiently long to afford another opportunity of observing what has repeatedly struck me in various parts of Germany, that is, the confiding honesty of the peasantry. Like all other orchards in Germany, the one above referred to was completely open to the road, and very seldom a native, even of the poorest class, is seen to take advantage of the occasion of thus improving the dry crust of which his slender *mittagessen* repast might consist. Nor could one fail to contrast this with the high walls surmounted by broken glass, the threatening notices, and constant surveillance by which such places in this country are protected, and not, as in "Vaterland," by the innate honesty of the people.

The baths of WEILBACH are situated on slightly rising ground, in the valley of the Maine, midway between Mayence and Frankfort, and about two miles from the river. There is nothing in the shape of a town or even village at Weilbach, which merely consists of the bathing establishment. This is divided into three separate pavilions, or large boarding houses distinguished by fanciful titles—the "Flora," "Victoria," &c.

As the season, which lasts only from the beginning of May to August, was over, we found all these establishments closed, we therefore proceeded to the central house, a large plain barrack-like building, where for some time we in vain endeavoured to effect an entrance. An immense mastiff, who, up to this time had quietly but pertinaciously followed us, as though distrusting our intentions, now seemed to have made up his mind that we were burglars, and first declaring war by an angry bark, forthwith charged, *à derriére*, Mr. E., who was then rapping at the door. Luckily the din aroused a sturdy looking woman, who came out wringing some nondescript garment fresh from the tub in her stalwart arms, and incontinently throwing the soapy habiliment into the face of her faithful hound "Ah, great villain, take that!" she cried, at which, evidently wounded by the unjust attack, the gallant beast slunk off, to Mr. E's great joy. We explained to the heroine of the wash tub, that although the season was over, we were particularly anxious to spend a day or two in Weilbach, but found that this was out of the question, as all the rooms were dismantled, and the house was in the possession of an army of painters and decorators. Moreover, it would be impossible to allow any one to sleep in the establishment without special permission of the " Grand-Ducal-bath-manager," and as that high functionary was then on leave for the winter, we agreed that we must see what we could during this day. While speaking to the woman her husband made his appearance, and, under his guidance, we proceeded to make ourselves acquainted with Weilbach.

The mineral source from which Weilbach derives its celebrity rises in a garden, a short distance to the right of the principal building. It is contained within an ornamental iron pavilion, on entering which, a very strong sulphurous odour is at once perceptible. The water flows into a marble basin in the centre of this, is very clear, of a slightly saline, alkaline, feebly sulphurous taste, but is not disagreeable, and has a temperature of 55 deg. The sides of the basin were thickly coated by a soft, whitish, soapy deposit, which consists of carbonate of lime, sulphur, and a peculiar organic matter. The water, though not sparkling, still contains nearly six cubic inches of carbonic acid, and three cubic inches of sulphurated hydrogen gases in the pint. According to the most recent analysis, the same quantity of water contains $4\frac{1}{2}$ grains of carbonate of soda, $2\frac{1}{4}$ grains of carbonate of lime, 2 grains of carbonate of magnesia, 2 grains of common salt, 1 grain of chloride of magnesia, a little sulphate of soda, and some other ingredients, amounting in all to $12\frac{1}{2}$ grains. From the spring the water is conveyed by leaden pipes into the large bath-house, which we proceeded next to visit. The interior of this building is as plain as the exterior, but is large and commodious; on the first floor are the dining, billiard, and reading rooms; and on the stories above are about a hundred plainly furnished, but comfortable rooms, the price of each of which was marked on the door, and varied from forty-eight kreutzers to three florins and a half. Were I myself sentenced to pass a season in Weilbach, I should certainly select one of the attics, as the vista from the upper windows is almost picturesque, while from the lower stories there is no view whatever. The other two buildings merely contain sleeping apartments, as all the guests breakfast and dine here; on the basement story are the baths, which are of the ordinary description, and I could find no trace of the mud-baths described by Mr. Wilson. The Weilbach water is principally used for drinking; but many of the patients conjoin the use of the baths. The great majority of the visitors are German, with some Russians; and as it is but little frequented by English invalids, a short account of its medicinal properties will be sufficient.

Used in full doses the Weilbach water is mildly aperient, in smaller quantities it is said to stimulate the appetite, and also to promote the removal of chronic visceral enlargements and congestions, especially of the liver. Most German physicians seem to be possessed by an extraordinary hallucination that all obscure diseases or complaints, the causes of which are not obvious to them (and I need hardly observe that the number is not limited), are produced by suppressed hemorrhoids; and it is further asserted by the local writers that the Weilbach waters are a panacea for these ailments. A large number of the invalids who visit this Spa suffer from thoracic affections, and we are told that incipient consumption, chronic cough, and spitting of blood are among the cases which come within the range of this remedy. I do not wish to express discredit of any particular writer; but I have no hesitation whatever in saying, as the result of very considerable experience of phthisis in various parts of the world, that I cannot believe that any consumptive patient was ever cured by the waters of Weilbach.

The next Spa in my itinerary was Soden, and thither I proceeded from Flörsheim by the Taunus railway, passing through luxuriant orchards, where the proprietors, gathering in the easy harvest seemed to be keeping a holiday rather than engaged in the toil of husbandry. Boys were perched in the trees rocking themselves to and fro, and with each shake a shower of golden cheeked apples came tumbling to the ground, while comfortably dressed, modest looking, and decidedly pretty peasant girls were filling sacks and baskets with the produce. Now and then one of them would fail to retreat in time, and be caught in a pelting shower of fruit, and great was the merriment of her more fortunate companions. I could not but contrast the scene before me with that probably enacting at the same moment in a country I had recently left; where gaunt, under-fed labourers, whose rags afford a poor protection from the chill rains of their climate, labour with reeking brow to draw from the ill-tilled land a scanty subsistence. Yet that soil is more fertile than this, and that peasantry more capable of toil, and more intelligent than the German "baues;" the one till their own little patrimony, while the other dare not make improvement in their cabin or farm for fear of thereby rendering their holdings more valuable, and of being consequently dispossessed without compensation. There are still, however, fortunately, Irish landlords who use the power confided to them with justice and humanity, and most of that class will probably agree that a power amounting in point of fact to very little less than that of life and death (for if the means of life be at the disposal of the arbitary will of an individual, life itself is at his mercy), should not be intrusted to any man, and that the relations of landlord and tenant should be adjusted on principles of equity to both. Such were my reflections when startled from my reverie by the hoarse whistle of the engine entering Höchst, and never did that sound more distinctly pourtray Sydney Smith's famous similitude of it to "the first yell of a roguish attorney entering Tartarus." From Höchst a branch line three miles long brought us to Soden.

Soden lies at the foot of the southern declivity of the Middle Taunus, below Königstein, nine miles south of Homberg, eighteen east of Wiesbaden, and only twenty-five minutes' drive by rail from Frankfort. The village—the old part of which is built in a little valley between the hills—is perfectly protected from the north and east winds by the Taunus range, while on the south it opens into the broad, sunny valley of the Maine. The modern quarter is built on the high road leading up the mountain; while several villas are perched on the hills above the old town. The hotels are numerous and tolerably good. The soil is peaty, and, as the only water found here is saline and mineral, the potable water is conducted from the neighbouring mountains. The saline waters issue from slaty rocks, and rise within a short distance of each other to the number of twenty-three.

The Soden springs are saline, acidulous, and ferruginous; they are clear, transparent, and gaseous, and the taste is more or less piquant and saline, according to the amount of carbonic acid gas and chloride of soda which the sources all contain in different proportions: from twenty-four grains, which is the minimum, in the "Milchbrunnen," to one hundred and twenty grains, which is the maximum in the "Salzquelle." The latter also contains the maximum of iron and carbonate of lime.

The action of the Soden waters are stimulant, diuretic, mildly aperient, and alterative. The resident physicians tell us that "the appetite invariably improves under the use of the water,—digestion certainly seems to be facilitated by the warm saline fluid, and this Spa appears to be particularly well adapted to the treatment of atonic dyspepsia, and languid digestion. The functions of the liver are excited by its use, and the biliary secretion is increased and becomes more healthy, therefore its use is indicated whenever this organ acts torpidly."

Dr. Sutro and other German writers advocate the exhibition of Soden water in "the tubercular diathesis, if used with whey, particularly in those cases complicated with anæmia. The extreme mildness of the climate here," he adds, "renders such a recommendation perfectly rational."*

The Soden spa is also, and I think much more judiciously, recommended in chronic mucous catarrh. This water moreover determines powerfully to the skin, giving rise in some cases to copious perspiration, and has been used with great success in some obstinate languid cutaneous eruptions, in scrofulous enlargement of the glands, and in chronic rheumatism.

* Dr. Sigismund Sutro's "Lectures on the German Mineral Waters," p. 217. London, 1851.

## WEILBACH AND SODEN.

I do not know a prettier spot among the Rhenish watering places than Soden. The view from the hill immediately above the village, across the broad valley of the Maine to the distant spires of the ancient imperial city of Frankfort, the pretty hamlets that are scattered through the plain so rich in waving corn and vine and fruit; the dark masses of the Taunus mountains behind, and nearer to the town, the slopes in which they begin, bright with verdure and foliage, interspersed with white fantastic villas, form a vista of quiet beauty, not surpassed in effect by many scenes of wider range and more striking features.

## CHAPTER XII.

#### HOMBURG AND NAUHEIM.

Frankfort—The Stadt Darmstadt—A night with Young Germany—The City of Frankfort—The Great Fair—HOMBURG-ON-THE-HILL—Its Topography—The Cursaal and its appropriate service—Number of visitors—The Springs—Their Doses—Physiological effects—Diseases in which they are resorted to—Gouty Dyspepsia—Hypochondriasis—Scrofulous Affections—Varicose Veins, Hysteria, and Diseases peculiar to Women—Duration of the Course—The Baths—Excursion to NAUHEIM—Description of this Watering-Place—Its Springs and their Medical Use.

AT Wiesbaden, a German confrere had informed us that we might find it difficult to get rooms in any of the large hotels of Frankfort, as the great fair was then going on, and had advised us to try the Darmstadt hotel. So on our arrival we accordingly drove to the Stadt Darmstadt, facing the entrance of the cathedral; but the house was not to our taste, being crowded, dirty, and noisy, and next day we removed to the Hotel de Russie, in the Zeil.

In justice to the Stadt Darmstadt, however, I should say, that the cookery was better than in many of the more fashionable houses, and the bill was moderate.

Frankfort is the head-quarters of German republicanism; and in this inn a club of patriots happened to hold their monthly meeting on the night we arrived. Thus we were afforded some opportunity of observing this phase of German character. On entering the dining room we found the walls covered with the rules of the club, patriotic songs, and prints, and over the head of the table was suspended a large daubed coloured engraving representing young Germany—not incumbered with any superfluous garments, crowning a statue of Liberty, whilst Justice, in a bright scarlet robe, was lashing out a multitude of kings and princes. Gradually the meeting collected, and about nine o'clock some twenty persons were assembled, most of whom were young men of the middle class, students, clerks, and shopkeepers. A plethoric little man, well dressed in black, now took the chair under the print I have described. Another who sat next to him then proclaimed silence, and opening a large book, read the minutes of the last meeting, which were signed by the president, who proceeded to address the assembly for upwards of an hour, but spoke so rapidly that I could catch only a few words till the peroration, when, forsaking prose, he burst into a song,—" Das Deutsche Vaterland," the chorus of which was sung by the whole club standing, and waving their pocket-handkerchiefs around their heads. A couple of similar speeches followed, succeeded by patriotic songs, some of which were certainly admirably sung. But, very soon, however, the party became more noisy and less musical. Whether it was grief for their country that made them so dry, or the singing, or the smoking, I know not, but certainly the patriots were thirsty souls, and consumed incredible quantities of bad beer and worse wine. In vain did the president now proclaim *silentium*; no one any longer paid the least attention to him; one portion of the club was chorussing a drinking song, the secretary, standing on his chair, was delivering an animated speech to the landlord, while the rest of the assembly were engaged in what appeared to us personal abuse of each other, so vehement were they, but which, in reality, was only a philosophical discussion. We had now had quite enough of this scene, and retired to our rooms, which, unfortunately for our repose, were immediately over the saloon where the patriotic conclave continued in full vigour till day-break.

Ancient, free, imperial city of Frankfort, mundane paradise of Jewish bankers, stock jobbers, German revolutionists, and Austrian soldiers—how heartily do I congratulate myself that thou art not a watering-place, and that it has not fallen to my lot to attempt to describe this most indescribable of cities! In one quarter we find a new city of wide and beautiful streets and palatial houses; in another we are transported back to the middle ages, not as in many cities, merely by the antiquity of the buildings, and the quaintness of the streets; but by the dress, appearance, and manners of the inhabitants.

Looking from the windows of the Hotel de Russie, on the Zeil, we might easily fancy that we were gazing on the boulevard de Sebastopol, the opposite houses have so much of the characteristics of modern Parisian architecture, fully as many soldiers are lounging about, and nearly half of the passers by are discoursing excellent French. Yet within five minutes' walk lies the Jewry, the most striking specimen of the customs as well as the architecture of the middle ages in Germany.

In one respect I was fortunate in my visits to Frankfort, namely, that I happened to be there each year during the great fair, and thus twice saw this most curious relic of mediæval commerce, which is still a mart of considerable importance, and by no means the holiday gathering of merry-makers gaping at jugglers and clowns, which some recent writers represent it.

Around the old cathedral lay, not a mere fair of tents and shows; but a regular city of booths, each street of which presented its peculiar speciality. In the principal street, if I may so call a line of wooden booths, the goods displayed were of the poorest description of English manufacture—cheap Manchester cottons, " Brummagem " jewellery, inferior Sheffield ware, and cast-iron cutlery from Leeds. In the next were costly philosophical instruments, and ingenious toys from Nuremberg, Bohemian cut glass, musical instruments, and wooden legs—yes, actually bona fide, wooden legs, to which one merchant confined himself exclusively, and of which he exhibited a stock which, I should have imagined, would more than supply all the hospitals of Europe for years. There were legs of all qualities and dimensions, legs ornamental, and legs plain, legs that would have supported a giant, and legs that would answer General Tom Thumb, should that gallant officer ever lose his understanding. I know not how to account for it, but this booth became to me an irresistible attraction; I wondered whether he ever sold any of these wooden limbs, and if so, who bought them. I tried in vain to realize to my mind a picture of the individual who came to a fair in pursuit of a wooden leg. Every day I felt myself impelled to visit this booth, but never saw any one in treaty for a leg. The proprietor seemed, however, in no way disheartened, but with a pleasant smile on his broad Teutonic face, sat constantly at the door of his warehouse with a pipe in his mouth, cheerfully polishing a superb ebony wooden leg. I got into conversation with him one day, and learnt more on the subject than I ever knew before, although I had assisted to make business for his trade.

Three-quarters of an hour's drive by the northern railway, brought us from Frankfort, to one of the most frequented Spas of Europe, HOMBURG-ON-THE-HILL. The little town, which lies nine miles north-west of Frankfort, and stands some three hundred feet above it, is situated on one of the lowest slopes of the Taunus range, immediately under the Great Feldberg peak. Of the three parallel avenues descending from the hill, of which, together with a couple of cross passages intersecting them, the town of Homburg consists, only one, the Luisen Strasse, deserves the name of a street. The upper part of this thoroughfare, with a few narrow lanes, forms the old town, round the castle; while the lower part, which is entirely modern, contains the railway station, cursaal, and hotels, forming the new town.

The Cursaal, occupies a prominent position in the main street, and is regarded by the inhabitants as the great feature of their town, being spoken of by them with that affectionate pride with which an Andalusian speaks of the Alhambra of Granada, or the mosque of Cordova, and in truth it is one of the most elegant dens of iniquity in all Germany. Herr Schick, a local historian, who devotes several pages to this venerable building, and treats of it with the most profound and unfeigned respect and gravity, tells us, that—" The first stone was laid after an *appropriate service*, and with the usual solemnities on the 23rd of May, 1841."* Unfortunately, we are not informed what the "*appropriate service*" consisted in, and so, are unkindly left ignorant of the proper form of invocation used by the assembled clergy, supplicating, at so much a head, the benediction of Heaven on an institution, to which the narrow-minded prejudices prevailing in this benighted country popularly affix the designation of a "hell."

The principal revenue of the state of Hesse-Homburg was derived from the rent of this establishment, the lease of which was renewed about a year before his death by the late and last Landgrave of Hesse Homburg, for an additional half-century.

* Schick's Guide to Homburg, and its Environs, p. 25. Homburg, 1859.

Some idea of the number of strangers frequenting Homburg may be gathered from the fact that, although almost every house is a lodging-house, there are about twenty-five hotels in this little town, of six thousand inhabitants. It is only within the last thirty years that Homburg has become a fashionable watering-place. In 1834 there were only one hundred and fifty spa-drinkers, while last season there were about twelve thousand visitors to the springs of Homburg.

The geological formation of the country round Homburg is chiefly a quartz slate; and the mineral springs issue from a thick vein of quartz, covered by strata of gravel and clay, lying a hundred and fifty feet below the surface.

The mineral sources of Homburg are all saline, the principal ingredient in the springs being common table salt, or chloride of sodium, in addition to which they contain a small proportion of iron. They are also all more or less acidulated, containing, for the most part, a large amount of carbonic acid gas. Besides these, however, they contain other constituents, the nature and amount of which will be detailed in speaking of the various sources. All the springs, however much they may differ in strength, and even in composition, rise in close proximity to each other in the Kurgarten, a beautifully kept park, connected with the cursaal, and occupying a small valley on the south-east side of the town.

I shall now proceed to describe the several springs,

in the order in which I examined them. The principal source is the Elisabethquelle, situated in a hollow at the extremity of an alley of poplars, and close to a covered gallery, in which the water drinkers can promenade in wet weather. This fountain was originally one of the sources of an abandoned brine manufactory, and had been neglected for upwards of a century, till its medicinal properties were discovered in 1834. The well, which is enclosed in a kind of paved court, is twelve feet in depth, and yields an abundant supply of pellucid sparkling water, the over-flow being estimated at nearly 12,000 quarts per diem. According to Baron Liebeg, a pound of the water contains seventy-nine grs. of common salt, ten grs. of carbonate of lime, seven grs. each of chlorides of magnesia and lime, two grs. of carbonate of magnesia, half a grain of carbonate of iron, and traces of several other salts, which it would be useless for practical purposes to enumerate here. The quantity of carbonic acid gas this water contains is its most remarkable characteristic, and in this respect it is one of the richest Spas in Europe, there being rather more than forty-eight cubic inches of gas in a pint of the water.

The Stahlbrunnen, which lies nearer to the cursaal, a little to the right of a long wooded avenue, is an artesian well, and was opened about twenty-five years ago. It is enclosed in an octagonal shaped space, in the centre of which the water rises through a perforated tin disc, placed about a foot under the surface of the well. The chemical composition of this spring is very similar to that of the Elisabethquelle, but it is not quite so rich in carbonic acid gas, and as the name implies contains somewhat more iron, amounting to nearly a grain to the pint. I thought the flavour of the water decidedly sulphurous, and my friend Mr. Egan made the same observation, but Liebeg's analysis does not indicate any ingredient which should occasion this taste.

The Kaiserbrunnen is also an artesian well, and issues from a glass basin. The very large volume of carbonic acid gas this water contains gives rise to a peculiar phenomena; viz., about once in every ninety seconds the water comes bubbling and simmering up in the basin, like water boiling on a quick fire, and then subsides into quiescence. This is evidently caused by the subaqueous accumulation and subsequent escape of gas. In each pint of this spa are dissolved one hundred and seventeen grains of common salt, thirteen grains of chloride of lime, seven grains of chloride of magnesia, half-a-grain of carbonate of protoxide of iron, a quarter of a grain of muriate of potash, together with forty-eight cubic inches of carbonic acid gas.

The next source is the Ludwigsbrunnen, which was the oldest used, and is the weakest of the Homburg springs. It is situated in a wood, in the Kurgarten, and issues clear as crystal from a glass basin in the centre of the fountain. Although this well was the first of these springs employed medicinally, yet it is no longer supplied by the same water as formerly, for in 1842, when boring for another well, the source of this was tapped, and it became necessary to deepen the shaft, which opened a new spring, differing materially from the former. According to Dr. Hoffman, a pound of the Ludwigsbrunnen water contains forty-three grains of common salt, seven grains of lime, one grain of potash, less than half-a-grain of iron, and forty-three cubic inches of carbonic acid gas. The temperature is forty-nine degrees, and the taste, although very saline, is to me agreeable.

The dose of Homburg water varies from six to forty-eight ounces, and must depend, in each particular case, upon the patient's age, sex, constitution, and ailment, as well as on the spring he may be advised to use; the Kaiserbrunnen, for instance, being double the strength of any of the others. In general, however, small draughts of eight or ten ounces, repeated at intervals of twenty minutes, are more advisable than a larger quantity at one dose.

Excepting in the quantity of their constituents, which vary so materially however, as to render the advice of a local practitioner indispensable to every one commencing a course of these springs, there is but little difference in the character of the various sources I have just described. All the mineral sources of Homburg belong to the class of ferro-saline waters; of which Kissingen, in Bavaria, and Cheltenham, in England, are examples, though of very different degrees of strength.

The physiological effect of the Homburg springs is chiefly produced by the amount of carbonic acid gas they contain, by which the respiration is quickened, the pulse is excited, and an exhilarating and agreeable sense of warmth in the stomach is experienced soon after the ingestion of the water. Its next action is generally that of a saline cathartic, but owing to the chalybeate salt dissolved in it, it seldom causes the weakness and subsequent ill effects often occasioned in debilitated patients by remedies of this class. The Homburg Spa, moreover, is capable of producing diuretic, alterative, resolvent, or anti-acid effects, according to the spring used, the dose in which it is taken, and the state of the patient.

The following are the principal maladies in which this spà is resorted to by British valetudinarians:— In gouty dyspepsia, attended by loss of appetite, furred tongue, painful digestion, acid stomach, and heartburn, derangement of the liver, torpid or irregular intestinal action, and depression of spirits,—a course of Homburg water often affords the best remedy, and sometimes leads to more permanent benefit than can be conferred by physic.

Hypochondriasis, which generally depends on gastric disorder, such as I have described, will, of course, be cured by whatever removes its cause, and this the

Homburg spa is often capable of effecting. It is also used in cases of general plethora, and in what the German spa physicians term "abdominal plethora," an affection to which they attach considerable importance, and, so far as the cathartic properties of the water go, I have no doubt that it is very properly employed in such cases; especially as the " spa doctors " moreover enjoin a regimen, which, of itself, would probably effect the cure.

The chloride of lime, contained by all these waters, increases the action of the absorbents and glandular system, and thus accounts for the benefits which have been occasionally observed to follow the use of this Spa, in cases of scrofulous inflammations of the glandular system.

In varicose veins and ulcers a course of Homburg waters is said to be productive of some advantage, by diminishing the tension of the diseased vessels, to which their chalybeate qualities impart a healthier tone.

In hysterical and nervous affections occurring in women, this spa is frequently serviceable, by removing obstructions, and facilitating functional discharges. In such cases it often proves the superiority of natural to artificial preparations, by the well-marked chalybeate action, which so small an amount of iron as that contained in these springs, the strongest of which holds only one grain of iron in a pint of water, is capable of exercising. For instances are recorded in which, after the pharmaceutical preparations of iron had been tried without success, a course of the Homburg springs has restored colour to the complexion, strength to the muscles, tone to the nerves, and crasis to the blood, in chlorosis and other similar complaints depending on obstructed functions and impoverished blood.

The usual duration of the "course" of Homburg waters is about three weeks, as if it be continued much longer than this, symptoms of what the spa-doctors term "saturation" manifest themselves. The patient now suddenly experiences a positive aversion to the water, and if, notwithstanding this, he still persist in its use, he will suffer from slight febrile disturbance, loss of sleep and of appetite, and rapid diminution of strength.

The mineral springs of Homburg should never be used in cases in which any organic disease exists whatever be suspected or apprehended, nor whenever there is any tendency to hæmorrhage; it should be especially shunned by patients suffering from pulmonary disease—in all such cases their employment may lead to dangerous, or even to fatal results. The waters are also, though not to any great extent, employed for bathing. The sources from which the baths are supplied are the Kaiser and the Ludwigs Brunnen, which, according to Dr. Müller, "combine the effects of the saline bath with those of the carbonic acid chalybeate one, and, consequently, while acting as a resolvent, they directly vivify and invigorate the system."* The large amount

* *Treatise on the use of the Mineral Waters of Homburg.* by Fredrick Muller, M.D., p. 71. Homburg, 1858.

of saline matter they contain, renders these waters, when used externally, highly stimulating to the skin. And Dr. Edwin Lee says, " as a revulsive measure against a congestive state of the abdominal circulation, evidenced by the occurrence of piles and other symptoms, as also in some cases of chronic eruptions without tendency to inflammation, of chronic rheumatism, in glandular enlargements, and other scrofulous affections, these baths may be advantageously used.* Dr. Prytherch, a resident physician, however, disapproves of these baths *in toto*, and says, " there is no fact of which I am more fully convinced than that of the inadmissibility of these stimulating warm baths simultaneously with a course of ferro-saline waters."†

The visitor to Homburg should not omit to see the neighbouring Spa of Nauheim, which is within an easy and picturesque drive of the former town. We started from Homburg in a carriage drawn by two strong, fast-trotting horses, and passing through the ancient town of Friedburg arrived in Nauheim at mid-day. Just before entering the town we drove by the remarkable " salines," or salt-works, which look like so many barricades defending Nauheim, and within five minutes alighted at the Cursaal.

NAUHEIM is situated on the declivity of the Johannisberg hill, on the railway between Frankfort and Cassel, and about an hour's drive by rail from the former.

---

\* *Homburg and its Mineral Waters*, p. 21. London, 1853.
† *Observations on the Mineral Waters of Homburg*, by J. H. Prytherch, M.D., p. 10. London, 1853.

The town is new and unfinished-looking, containing a population of about two thousand inhabitants. The streets are laid out in wide and pretentious-looking boulevards, which are quite unsheltered by the newly-planted trees; the houses are built on the French model, and are scattered along these, facing the park, in short there is a look about Nauheim as if spasmodic efforts had been made to convert it into a fashionable watering-place, and had signally failed. The Cursaal, a large, tawdry specimen of fantastic architecture, was opened since I had last visited the Rhenish Spas, and great efforts have been made to make it a rival to Homburg or Baden. The rooms, however, were nearly empty when we entered, the roulette table was abandoned, and the croupiers sat idly round it rake in hand. Nor was the rouge-et-noir table doing much better, only two or three persons were playing for small silver florins. On the terrace outside an excellent band was performing to half-a-dozen listeners, so turning away from the deserted gambling house we proceeded to visit the springs. And I am indebted to my very observant companion for some of the following notes on the saline springs of Nauheim.

These waters have long been employed in the manufacture of salt, but only within the last few years have been resorted to medicinally. There were formerly a vast number of natural saline springs here, but their sources were gradually interfered with by the sinking of artesian wells, of which there are six or seven. The principal of these is the "Grosser-Sprudel;" one of the most remarkable springs in Europe. This sprudel burst from an unfinished artesian well, in the night of December 21st, 1846, and spouted out in a thick column of water to a height of nearly eighteen feet from the surface, and thus it has continued to issue ever since. It is enclosed within a large open stone basin, the water in which from the vast quantity of carbonic acid gas it contains is white with foam. It contains, besides, two hundred and fifteen grains of saline matter to the pint of water, of this one hundred and eighty-one grains is common salt, together with fourteen grains of chloride of calcium, eleven grains of carbonate of lime, four grains of chloride of potassium, with other salts, and a trace of bromine. Its temperature is not easily ascertained with exactness, but is about ninety degrees. This source is strongly purgative in doses of one glass, but is principally used for bathing. The baths are adjacent on each side of the spring, and are used at the natural temperature of the water.

The Kurbrunnen and Salzbrunnen rise near each other. The former is that usually employed internally; it is the weakest of all the Nauheim springs, containing a hundred and thirty-three grains of salt to the pint; in the chemical composition of its ingredients it, however, resembles the Grosser-Sprudel, as also does the Salzbrunnen, which contains a hundred and sixty-nine grains, and the Kleiner-Sprudel, containing one hundred and eighty-one grains, in the same quantity ef water.

The effect of these springs used internally is purgative and diuretic. Their principal use is, however, for baths, in cases requiring a stimulating application to the skin, in obstinate and languid cutaneous complaints. They are said to exercise a specific action in accelerating and increasing the catemenia. They are also prescribed in certain cases of chronic rheumatism, in scrofulous tumours, and diseases of bone.

## CHAPTER XIII.

KISSINGEN, BOCKLET, AND BRUCKENAU.

Early rising—Midnight review of a sensation novel—Departure from Frankfort—The vineyards of the Maine—Bavarian railways—A Turkish missionary—WURZBURG—Schweinfurth—Mountain drive—Bavarian politeness—KISSENGEN—Description of the town—The mineral springs—The Spoolenbrunnen—Remarkable phenomenon—Medicinal qualities of the waters—Visit to BOCKLET and BRUCKENAU—Account of these watering places—Their sources and their properties.

We had loitered somewhat too long on the borders of the Rhine, and amid the "Brunnens" of Nassau, and now began to be conscious that the season was drawing on apace, while we had as yet done little more than commence our pilgrimage to the Spas. So having returned to Frankfort, we came one night to the determination of visiting the Bavarian watering places before going on to our destination in Bohemia. No sooner had we made up our minds than we summoned the "Ober-kellner," settled our bill, and requested to be called next morning at the terrible hour of half-past three a.m. As we repeated this injunction, something very like a sardonic grin passed across the countenance of this best trained of waiters; he evidently thought we meant to be facetious, but seeing that we were in sober earnest, he instantly assumed a becoming and decorous look of melancholy sympathy at the prospect before us, and promising that we should be called, retired, leaving us to our fate.

My fellow traveller had wisely made his preparations beforehand, and so escaped the misery of packing, to which I devoted a part of the night. After some hours of ineffectual labour to restore my effects to the order in which they had been when I had extracted them, every article neatly folded, laid in its proper place, and easily accessible, finding they obstinately resisted every gentle effort to replace them, in despair I gathered my scattered property together, and forced a chaotic heap of garments of all descriptions—books, crumpled coats, engravings, bottles of eau de cologne, surgical instruments, and geological specimens, into my portmanteau; which I succeeded in closing, with the aid of the stout night porter, whom I brought from his post in the hall to my assistance. It was now two o'clock a.m., and, as it seemed useless to go to bed, I took up the only book at hand, one of Miss B.'s sensation novels; and owing, perhaps, to my want of acquaintance with this class of literature, it struck me as one of the most unpleasing works I ever read—a constant repetition of startling effects, and impossible situations, well tinged with mystery, and highly spiced with bigamy and murder, abounding in slang phraseology, and very questionable morality, yet withal so cleverly told, and so interesting, that I found it impossible to lay down the volume, till a knock at the door apprized me that it was time to start.

We left Frankfort by the Royal Bavarian Railway, and, shortly after our departure, saw by the white and blue stripes on the signal posts, that we had entered the Bavarian territory. For the first ten miles, as far as Hanau, the country was flat and uninteresting; but after leaving that station, we first passed through a wide-spread, rich-looking praire, and then, beyond Aschaffenburg, entered the sole remaining portion of the vast Hercynian Forest, which, according to Cæsar, once covered a great part of Germany. Emerging from this, the line runs along the valley of the Maine, lying between the river and the long parallel range of the Spessart mountains, the lower slopes of which are cut into terraces for the vineyards, which covered the entire country for the remainder of our journey to Wurzburg.

Apropos of vineyards, I find the following jottings set down in my note book as we journeyed from Aschaffenburg to Würzburg, and, as I believe they are tolerably true, I have transferred them just as they were. There are several fallacies prevalent about the German vineyards. The first is that the vineyard is an ornamental crop. We read a great deal about " clustering vines," &c., but having journeyed through almost every vine-producing country, and resided for years in several of them, namely, in Portugal, Spain, and France, I must say that, excepting in Tuscany, where nothing can exceed the beauty of the festooned vines, the grape is not the poetic-looking crop that it is generally supposed to be by those who have never seen it.

The French vineyards are no exception to this rule. Nothing can be less pleasing to the eye than the interminable undivided tracts of yellow clay, planted with little scrubby bushes, supported on long stakes, which form the great bulk of the French vineyards, at least in the northern departments. Something similar to these, though certainly superior in beauty, are the Spanish and Portuguese vine plantations, and those of the South of France. But the German vineyards, especially on the banks of the Rhine, are, *par excellence*, the ugliest in Europe, and yet every cockney tourist complacently apostrophises, "This vale of vintage bowers." The said " vintage bowers" being represented by a series of terraces as wide as an ordinary door-step, supported on rubble masonry, built into the sides of perfectly barren hills, and covered with earth, diligently conveyed hither from some kinder soil. On each of these steps are planted a couple of dozen little bushes, the aggregate produce of which, throughout the entire Rheingau, would hardly furnish the quantity of so-called Hock, Johannisberg, or Steinberg, used in London on a summer's day.

These Bavarian vineyards which we were now passing through, were however, very different, and covered the whole country, from the winding banks of the

Maine to the distant mountains, with a luxuriant harvest of large purple grapes, trellised as in Italy, from long poles, presenting very much the appearance of the hop-gardens of Kent. That we were now in a Catholic country was evinced by the little statues of the patron saints of the vineyards, placed as in Spain, in the centre of each field.

All along this line of railway we noticed a remarkable neatness in the station-houses, which were covered with creepers and flowers, nor was there a single inch of waste land visible; every embankment was cultivated, and even the huts of the signalmen were surrounded with little flower gardens. The carriages, though not equal to those on the Nassau railways, were superior to those on any line in this country, and the officials were becomingly dressed in a kind of military uniform of a light blue colour. There was, however, no great appearance of traffic, and our two fellow-passengers, a Scotch gentleman with his daughter, and ourselves, were left the whole day in undisturbed possession of our coupé. This gentleman, like most Scotchmen, was a shrewd and intelligent observer of men and manners, possessed of a considerable amount of dry humour, and yet was strongly imbued with the religious prejudices peculiar to his country. He had spent seventeen years as a missionary in Turkey, and was now on his journey back to Constantinople, where he was settled. He seemed full of hope for his labour, and said that, though as yet they had made few, if any, converts, they had great expectation of reaping the harvest they had so long toiled at. I fear I must have shocked this good man by my opinion, founded on what I have seen of the Turkish Empire—that the Turks are utterly and irretrievably an unimproveable people. At Würzburg, however, we parted company, the missionary having a weary journey before him to Vienna.

We had an hour's delay here, of which we availed ourselves to drive down to the town and get a view of the cathedral. From the flying glimpse we thus had of Würzburg it struck us as a very sombre, dilapidated looking, ancient town, that had long since seen its best days, and was now in mourning for the decadence of its fortunes. Hardly ever did I see so many churches collected together as in this city: on every side rose time-honoured domes and venerable spires, the monuments of the prince-bishops, who so long ruled this quondam capital.

Hurrying back to the station, we had barely time to refresh ourselves with a flask of the celebrated Steinwein, grown in the neighbourhood of this city, before our train started for Schweinfurth, where, in another hour we arrived, and driving to the Hotel Rabe, found ourselves just in time for dinner.

The great event of the day over, and having ordered a carriage to take us on to Kissingen at six o'clock next morning, we strolled out to see the town, and returned back in ten minutes, having accomplished this undertaking. For, whatever Schweinfurth may have been in the middle ages, when it was an Imperial City, and the great corn emporium of Germany, it is now merely a large uninteresting village, that looks as misplaced within the ancient civic walls which still encircle it, as does the hut of a Granadian gipsy in one of the vast roofless halls of Charles the Fifth's ruined, yet unfinished palace, in the Alhambra.

Next morning our conveyance was at the door before we had shaken off the chains of Somnus; and so, scalding ourselves with the boiling coffee, we hurried down and found that our civil landlord had put his own carriage at our service, which looked more suited for Hyde Park than for the rough mountain road before us. A prettier drive than that from Schweinfurth to Kissingen could not be readily met; though so hilly was it that, for the chief part of the way, we were obliged to drive at a foot pace. One little village we passed through was gaily decorated with flowers, and at the end of the street was a very pretty rustic triumphal arch, surmounted by a bishop's mitre. What the occasion of this festivity was we could not ascertain, nor was there any appearance of merry making in the street. Half-a-mile further we passed a train of boys and girls returning towards the village, the girls all wearing chaplets of flowers, and dressed in a costume very like that which I had supposed was confined to the traditional stage peasant of our theatres, but which, as I now saw, has a real existence. Each of the lads took off his hat and saluted us as we passed—and indeed this politeness seems universal in Bavaria among every class. Even the coachmen of the various carriages we met returning from Kissingen, instead of " chaffing," or even imprecating each other, as I have more than once seen done under similar circumstances in this country, exchanged grave and formal salutations with our driver, the leaf of whose hat had acquired a rakish bend over his right eye, from the frequency with which it was lifted up on that side.

After a drive of nearly three hours we reached Kissingen, and now, after the fashion of all coachmen, our driver, anxious to wind up with a grand flourish, galloped hard through the town, and was just turning into the court yard of the " Hotel de Saxe," when the old adage of " pride will have a fall," was verified by the front wheels suddenly becoming locked, and our headlong impetus being thus violently checked, the carriage slowly rolled over on its side. We were luckily unhurt, and scrambled out through the opposite door, and the carriage, only a little scratched by the accident, was soon set to rights. The only sufferer by the mishap was the driver, who seemed profoundly hurt in mind, though not in body, by the blow his reputation as a whip had sustained. We tried to comfort him but did not succeed, though his dejection was somewhat removed by a liberal " trinkgeld."

Kissingen, in the department of the Lower Maine, in

Bavaria, is situated about six hundred feet above the sea, on the Saale. The town stands in the centre of an extensive valley running north and south. This valley is surrounded by hills covered with woods, orchards, and vineyards, and through it the winding Saale makes its way southwards to join the Maine. The town itself is a mere watering place, composed entirely of large, newly-built hotels and lodging houses, which during the season generally accommodate about five thousand visitors; together with a few small shops, and needs no further description. Besides these, during the fashionable months, which extend from the beginning of June to the middle of August, there is a small theatre in the Bellevue Gardens on the Bruckenau Road, open from four till six in the afternoon.

The three principal springs rise close to each other in the "Kurgarten," a kind of small park, a little to the south of the town, between the street and the cursaal. The spring nearest to the entrance of the garden, and the one we examined first, was the "Maxbrunnen," situated in an oval-shaped paved court, above five feet below the surface of the Kurgarten. This spring, which is the only one not under cover, is left purposely open, from some idea that its medicinal effects would be impaired were it not exposed to the sun's rays. In this space there are two deep wells, only one of which is used. The water rises from the sandstone rock with a loud hissing noise, and is bright, clear, and sparkling, containing thirty-one cubic inches of very finely divided carbonic acid gas in the pint. This source possesses very little taste, but is rather acidulous and agreeable, and contains about thirty grains of saline ingredients in a pint, the principal constituent being common table salt, of which it contains eighteen and a quarter grains. The temperature is 52 deg. While we were examining the spring, a number of working people came up, to whom the attendant handed glassfuls of it without any charge, and many of them carried away bottles and jars of this water, which is used as the common drink of the inhabitants, and is said to be the cause of their immunity from scrofula and parasitic complaints. It stands to reason, however, I think, that a fluid which acts as a powerful remedy in disease cannot be a proper daily beverage.

The Pandur and Ragoczy are situated a few yards nearer the cursaal than the last spring. They are contained in the same sunken area, which is about sixty feet in length, and is covered by a very ornamental light iron pavilion. The Pandur is the first seen on entering, and rises from a well about twenty feet deep. The temperature of the water is 52 deg. It is not quite so gaseous as the Maxbrunnen, but is much stronger in taste, being very saline, and somewhat chalybeate. This spring contains more than double the amount of salts of the last described, or seventy-six grains of solid ingredients to the pint. It is chiefly employed for baths, but may be drank in many cases in which the Ragoczy is too powerful. Like that spring, its action is aperient and solvent, but it moreover increases the cutaneous and renal, as well as the alvine secretions.

A few yards to the right of the last described spring is situated the most celebrated of these sources, and that which is generally understood when " Kissingen Water" is spoken of, namely, the Ragoczy, or Rakoczy as some call it. This spring was discovered in 1738, in the old bed of the river, which was then turned into its present channel. The water rises through a pebbly bottom of basaltic sandstone into a well about twelve feet deep. It is bright and colourless at first, but after being some time in the glass becomes discoloured, and deposits a brownish precipitate. Its taste is extremely unpleasant, being saline, bitter, acidulous, and somewhat chalybeate or astringent. Its composition and strength sufficiently account for its marked and compound taste; a pint of the water containing no less than eighty-five grains of salts, of which quantity chloride of sodium alone amounts to sixty-five grains, together with which there are sulphates of soda and lime, chloride of calcium and magnesia, some bromides, and other salts. It is used only for drinking, and I need not dwell on its effects here, as what I have to say hereafter on the medicinal qualities of Kissingen water, refers principally to this source.

About a mile from the other springs, to the north of the town, rises the Spoolensprudel. This well is contained within a handsome square building facing the river. The hall occupies the entire central part of the structure, and is lighted from the dome, while at each side different galleries run round. In the middle of this apartment is a raised structure, somewhat like an immense bee hive, with glazed cover, about two feet high, beneath which is the Spoolensprudel. The shaft of the well seen through this cover is about twenty feet deep, by about eight feet in diameter, and seems to be lined with woodwork. Through the bottom of that shaft an artesian well was bored in 1822, to the depth of nearly three hundred feet, in the sandstone, and this it is which now supplies the Spoolensprudel.

This source presents a very remarkable phenomenon, namely, it intermits, or ebbs and flows, with great regularity, eight or nine times daily. If we arrive a little before the "flow," on looking down the shaft through the thick glass cap, we see the well apparently empty of water; we then hear a distant rumbling noise, which gradually becomes louder, and draws nearer, till in about half-an-hour from the time it was first heard, the water is seen foaming below; it gradually ascends, and in another quarter of an hour reaches within a few inches of the top of the well, covered with white foam, and hissing and seething with great turbulence. Above the water hangs a heavy layer of carbonic acid gas, rising and falling with it. This gas is collected and ingeniously utilized in various forms of gas baths. Indeed, the entire system of vapour, gas, and water baths in the establishment is extremely perfect.

Internally used this water is strongly purgative, and,

diluted, as from its strength it could not otherwise be drank, it is employed in the same class of cases as the other springs. It is, however, principally used for bathing, in rheumatic, neuralgic, and cutaneous affections. It acts as a powerful stimulant to the skin, producing redness, and considerable irritation of the surface. From its close resemblance to sea water, having however, in addition, a small proportion of iron, and a considerable amount of carbonic acid gas, it might, and perhaps even more advantageously than sea water, be employed in scrofulous cases. The mother lye, or concentrated saline water of the Spoolensprudel, is applied with wonderful result as a local application to scrofulous glandular swellings, and similar affections. The following is the analysis of the foregoing sources:—

COMPOSITION OF KISSINGEN MINERAL SPRINGS.

|  | Ragoczy. | Pandur. | Max. Brunnen. | Theresen Brunnen. | Spoolen sprudel. |
| --- | --- | --- | --- | --- | --- |
| Carbonic acid (cubic in.) | 26·25 | 28·85 | 31·04 | 28·35 | 30·57 |
| Chloride of sodium | 65·05 | 57·00 | 18·27 | 18·40 | 107·51 |
| ,, potassium | 0·91 | 0·25 | 1·00 | 0·85 | 0·97 |
| ,, calcium | — | — | — | — | 3·99 |
| ,, magnesia | 6·85 | 5·85 | 3·10 | 2·75 | 24·51 |
| Carbonate of soda | 0·82 | 0·03 | 0·38 | 0·39 | — |
| ,, lime | 3·55 | 5·85 | 2·59 | 2·00 | 1·65 |
| ,, magnesia | 2·50 | 1·62 | 1·82 | 2·37 | 6·41 |
| ,, iron | 0·68 | 0·45 | — | — | 0·35 |
| Bromide of sodium | — | — | — | 0·07 | — |
| ,, magnesia | 0·70 | 0·68 | — | — | 0·06 |
| Sulphate of soda | 2·00 | 1·75 | 1·86 | 1·35 | 25·30 |
| ,, lime | 2·50 | 0·75 | 0·65 | 0·75 | — |
| Phosphate of soda | 0·17 | 0·05 | 0·12 | — | — |
| Silica | 2·25 | 1·55 | 0·46 | 0·50 | — |
| Oxide of aluminum | 0·18 | 0·05 | — | — | — |
| Organic extract | 0·15 | 0·09 | — | — | 0·86 |
| Loss | 0·38 | 0·37 | 0·38 | | |
| Total solid contents in sixteen ounces | 85·74 | 76·39 | 30·65 | 29·63 | 187·68 |

The Kissingen waters are used in almost every variety of dyspepsia. The Spa physicians assert that they are the best remedy for chronic or habitual constipation, and, indeed, they certainly act as very brisk and active carthatics, leaving, too, no subsequent debility after their operation, as most medicines of that class do. This is probably owing to the tonic effects of the soluable carbonate of iron which they contain. Hypochondriasis, so intimately connected with irregular gastric or intestinal action, is therefore said to be peculiarly under the influence of the curative action of these waters; and judging from the physiognomy of the visitors, half the invalids who drink the Ragoczy seem to labour under that malady.

In chronic enlargements of the liver, and also in passive congestion of that organ, Kissingen water is often used with considerable advantage, whether the enlargement and hardness result from the effects of a tropical climate, or from high living, and the excessive use of stimulants. In such cases the spring generally used is the Ragoczy, and it must usually be persevered in, even after the patient having gone through the ordinary course at Kissingen has returned home. Enlargement of an adjoining organ—the spleen, which so frequently results from the intermittent fever of hot countries, and which my experience in them leads me to consider a most difficult disease to cure by the remedies commonly employed, namely, mercury and quinine, is sometimes completely dissipated by a course of this

Spa. The German physicians are wont to recommend a course of the Maxbrunnen internally, conjointly with the Pandur externally in baths as a preventive of scrofula in children predisposed to that disease, and also in the treatment of scrofulous glandular affections. This practice seems to be well founded, for not only does scrofulous disease often yield to the Kissingen Spa, but, moreover, the inhabitants who use the Maxbrunnen diathetically are, and as it is asserted on that account, almost exempt from scrofula, which is a very rare affection with the residents here.

The Ragoczy is prescribed in cases of irregular gout, but I certainly could never think of sending any gouty patient of mine to Kissingen. For this Spa is a stimulating carthatic, and as Sydenham very truly remarks—"Nature diverted from her own good and safe manner of depositing the peccant matter in the joints, as soon as the humours are solicited towards the intestines, instead of the acute pains with little danger, induces sickness, griping, fainting, and other irregular symptoms, which will nearly destroy the patient."*

In cutaneous affections, when merely symptomatic of gastric derangement, these springs are considered to act as specifics. I know of hardly any mineral water which is not said to possess the property of curing sterility, and perhaps there is no place which has attracted so much attention on this account as

* Sydenham, on the Gout, Sec. XXII. Works, p. 131.

Kissingen. But although Dr. Granville vouches for some extraordinary instances of its efficacy in this way, yet considering that the causes of this condition are so manifold, and for the most part so obscure, I doubt very much whether any mineral water can do much more than remove some of these causes, and when these are known, they may probably be better treated by other remedies than the waters of Kissingen.

Nearly five miles from Kissingen is the Chalybeate Spa of BOCKLET: the road to which passes the saline spring I have just described. Bocklet is a little village of lodging houses grouped around a central "Conversation-Saal," or bath-house. There are here several mineral springs, which are used internally and for bathing. All the sources are cold, chalybeate, and very gaseous; they rise close together, from a red sandstone formation, and are enclosed in a handsome pavilion on a square terrace. These saline-ferruginous waters contain about two-thirds of a grain of iron, together with twenty-seven grains of chloride of sodium, seven grains each of carbonate of lime and sulphate of soda, and other salts, amounting in all to nearly forty grains of saline matter to the pint. The most important constituent of these waters, however, is carbonic acid gas, of which they contain thirty cubic inches to the pint; and it is this gas which confers its peculiar activity on the iron dissolved in the spa. As may be supposed from its composition, the Bocklet mineral water is one of the most powerful saline chalybeates in Europe, and is peculiarly adapted to cases in which weakness of the digestive system is a prominent symptom.

From Kissingen a picturesque mountain drive of about fifteen miles leads the spa-tourist to the important watering-place of Brückenau, which lies in the beautiful wooded valley of the Sinn. The springs are between two and three miles from the town, and around them are situated the baths, hotels, and lodging-houses, the royal palace, and Kurhaus. Two of the springs are on the left bank of the river, and contain only a small quantity of the salts of soda. One of these, the Sinnbergerquelle, is said to possess diaphoretic properties, and is also prescribed in chronic bronchitis, scrofula, and calculous affections. On the opposite bank of the river is the Brückenau Quelle, from which this place derives its reputation as a chalybeate spa. It rises under a pavilion, and flows into a basin incrusted with protoxide of iron. The taste of the water is strongly ferruginous, although it only contains a quarter of a grain of iron to the pint. But this is rendered most active by the immense quantity of carbonic acid gas, near thirty-eight cubic inches of which is combined with it.

The medicinal effect of Brückenau Spa is tonic, and very stimulant. Its use is, therefore, dangerous whenever there is any organic disease; or even when functional disease, in which a powerful excitant would be improper, is present. It is, however, a very valuable stimulating chalybeate in certain cases, which I have already pointed out in the introduction, and need not repeat here.

## CHAPTER XIV.

#### THE JOURNEY TO CARLSBAD.

Kissingen to Bamberg—A Bavarian buffet—An apparition—Hof—Journey to Zwickau—Schwartzenberg—Disinterested benevolence—Drive to Carlsbad—Austrian frontier—Attempt at robbery—Arrival in CARLSBAD.

Misled by that nondescript compilation, so incomprehensible, and yet so indispensable to tourists—Bradshaw's Guide, we left Schweinfurth by the mail train in the evening, en route for Carlsbad, which, however, we were not destined to arrive at as soon as we expected. In a couple of hours we had reached Bamberg, where we had a long delay. In the meantime we followed our fellow passengers into the waiting room, which, as is usual in Bavaria, is also the buffet. Here we found the room crowded to excess with travellers of every class, most of whom were smoking, struggling to reach the counter by dint of sheer fighting, such as I never saw equalled in civilized life, except at a ball by some ardent youth fighting his way through the throng, with a semifluid ice, to the object of his admiration.

Escaping from the stifling atmosphere of this apartment, I was sauntering up and down the platform, when a counterpart of the beadle from the Burlington Arcade presented himself before me. There was the same gold-laced hat, the same graceful coat flowing down to his heels, the same long official staff with its immense ferula, and silver globe on the top; and, above all, there was the same slow, majestic strut, and awe-inspiring demeanour, which none but a beadle could assume, and none so well as him of the Burlington Arcade. But what could possibly bring him to Bamberg at midnight in his official costume? I reasoned—

"I think it is the weakness of mine eyes that shapes this monstrous apparition."

The spectre, however, soon solved my doubts as to his character, for, stationing himself at the door of the Restauration-saal, he made a brief but impressive speech, announcing the arrival of the trains for Hof, Nuremberg, Augsburg, and Munich.

Our train now came up, and taking our seats in an empty compartment, we were just falling comfortably asleep, when, at the first station, three ladies and an infant were introduced, and I need not say that sleep was now out of the question. The child squalled incessantly, and the ladies talked in that loud, piercing voice which most German women have, whenever they were not eating, and whenever they stopped talking till they arrived at Hof.

We reached Hof at half-past four a.m., and then learnt that the coach for Carlsbad had left half-an-hour previously, and that there would be no other conveyance until next morning; but if we went on to Schwartzenberg we would there find a diligence for the Bohemian Spa. While we were picking up this information however, the train for Schwartzenberg started, and having now six hours to wait for the next, we strolled into the old frontier town of Hof, and early as it was, found most of the inhabitants stirring. The Germans are certainly the earliest risers in Europe; for where else but there would we find the shops of a large town open, the markets going on, and the people pursuing their avocations before five o'clock in the morning.

We must not pause, however, to describe Hof, which is more of a Saxon than a Bavarian town, though it partakes of the peculiarities of both countries, being Lutheran like the first, and beer-drinking like the latter.

From Hof to Werdau we passed through a quiet pastoral country, not unlike the midland counties of England, now and then varied by small manufacturing towns. Beyond Zwickau the character of the scenery changes completely, and becomes of singular beauty; the railway winding along the banks of a small river, through exquisitely wooded glens and hills, where the course of the line was the most winding I ever saw.

Six hours after our departure from Hof we arrived

at Schwarzenberg, where we found that we had had our journey for nothing, as the diligence for Carlsbad had discontinued running the previous day. It was now too late to hire a carriage to take us on, therefore leaving our luggage at the station, we proceeded in search of an hotel, which, in this old-fashioned, little-frequented place, we had some difficulty in procuring. We tried two or three inns near the railway station, but found them all dirty and comfortless, and were just turning away from the Hotel de Saxe, when a well-dressed individual, with a perpetual smile on his greasy countenance, accosted us in some unknown tongue which I fancied, but could not be positive, was intended for French. After several efforts to make himself understood, he consented to speak German, and said he could conduct us to " a first-rate hotel, an hotel worthy of Dresden, an admirable hotel." He grew enthusiastic in praising this wonderful hotel, he kissed the tips of his fingers and turned up the whites of his eyes in ecstasy at the very thought, and wished by main force to carry up our carpet bags; but we could not think of allowing him to trouble himself so far. At length, after a walk of nearly a mile up-hill towards the town, we arrived at the hotel in question, which was a dirty-looking inn, of the poorest description. As we declined to enter, our friend waxed exceedingly indignant. " Was that the way to treat a gentleman?" he asked. " Did we wish to insult his hotel?" and several other similar inquiries, of which we took no notice. Entering the old town, which is built round an extensive fortified castle on the summit of the hill, we lighted on a large " Gasthof" in the " Platz," where we obtained a couple of very comfortable rooms and a tolerable dinner, with a bottle of old Johannisberg.

Next day the carriage, which we had ordered for noon, was not ready till near four o'clock in the evening, when we started. The road soon became very hilly, passing through a thinly inhabited wild mountainous country, not unlike that between Bray and Blessington in the Wicklow hills, but on a much grander scale. We passed few houses excepting those at the turnpikes, which are numerous, and consist of a long pole stretching across the road, which is lifted up by an invisible toll man, who from his box extends something like a fishing rod, with a bag fastened to the end, into the carriage, and thus collects his dues.

A little after dark we arrived at the Austrian frontier, where, for the first time since we left England, our passports were examined. The custom-house officer, who addressed us in excellent French, was, as is invariably the case in Austria, extremely courteous and obliging; while in the country we had just left, the Saxon officials are as constantly rude and troublesome. On the present occasion we were merely asked " Had we anything contraband?" and on replying in the negative, our luggage was not examined.

An hour later we again stopped, and on looking

out, saw that the horses were being unharnessed. The coachman next proceeded to put out his lamps, and went off without any explanation. We descended and found ourselves in a large unlighted village, drawn up at the door of a public-house, where, on looking in, we saw our driver, with several others, seated at supper. Returning to our carriage, we waited patiently for upwards of an hour, till hearing a noise immediately behind us, I got out, and found three or four lads assembled behind the carriage, one of whom was undoing the cords which held our portmanteaus. I intimated to him that I disapproved of his proceedings by an argument addressed to his tenderest feelings. He took the hint and decamped with his companions. I now went in search of our coachman, who could not be found, till Mr. E. by dint of perseverance, at length dragged him out of the loft where he was sleeping off the extra potations of beer he had evidently indulged in, and by eleven o'clock we were once more *en route*.

The night was pitch dark; it rained heavily; the horses were tired; the roads were heavy, and the coachman was half asleep; and under these circumstances it cannot be wondered at that our progress was not the most rapid. We jolted on uneasily through the night, occasionally rousing up the driver when he seemed to nod, and were heartily tired when, about four o'clock in the morning, we entered Carlsbad, and at length alighted in the court yard of the Hotel de Russie.

## CHAPTER XV.

#### CARLSBAD.

Its situation—The warm river—Description of the town—Amusements—Hotels—Climate—Analysis of the "Kurliste"—The springs—History of their discovery—Account of the Brunnens—Action of the waters—Author's experience—A remarkable case—The Sprudel versus Mr. Banting in obesity—Chronic hepatic disease—Irregular gout and rheumatism—Dyspepsia and hypochondriasis—Cases for which Carlsbad is unsuited.

CARLSBAD, in Bohemia, is situated upon the river Teple, in a small deep valley, formed by the Erzgebrige Mountains. This valley stands about a thousand feet above the level of the sea, and lies eighty miles west of Prague, and nearly seventy miles south of Teplitz. The Teple, which flows through Carlsbad, owes its name to a Bohemian word, signifying "the hot one," and after passing through the town, at least merits this name; for I found the temperature of the river, before the overflow from the Sprudel and other springs falls into it, is but 56 deg.; whilst immediately below this point the temperature rises to 72 deg. Strange to say this warmth appears to be favourable to the fish, for I observed a quantity of remarkably large carp just below the junction of the warm water.

Carlsbad is one of the most beautifully-situated watering places in Germany. It lies at the bottom of a deep ravine, intersected by the Teple, and surrounded by lofty hills, wooded to their summits. From these, spurs project into the town, dividing it almost completely into two small triangular valleys. In the first is situated the chief part of Carlsbad, the hotels and all the mineral springs, while the second, through which the road to Marienbad leaves the town, contains merely a few scattered houses.

The resident population of Carlsbad does not probably exceed four thousand inhabitants, and the place may be described as consisting of a couple of streets called "Wiese," built along each side of the Teple, and connected by eight or ten narrow bridges.

The "Hotel de Russie," in which we lodged, is situated on the right bank of the river. On the same side are most of the other hotels, and the Sprudel; rising immediately above which there is a large handsome church. Behind these buildings there are a number of narrow streets of steps, somewhat like those of Malta, ascending the hill; and further on, close to the river side there is a small theatre.

On the opposite bank is the Alte Wiese, a long range of small shops, extending southward along the Teple. Behind this is the Hirschen-Sprüng, whence, as the tradition asserts, a stag, hunted by Charles IV., sprang across the river and fell into the Sprudel, which is about half a mile distant, and by its piteous cries in the boiling water, attracted the Emperor's attention, who, coming up, despatched the deer, and then refreshed himself with a warm bath. This had the effect of curing a bad leg, from which he suffered. And thus was the Sprudel discovered, and its virtues at the same moment proved beyond a doubt. This event was commemorated by order of the grateful monarch, by the name Carlsbad (or Charles's bath) being then affixed to the watering place, which at once sprang up around the Emperor's bath.

Further north, on the same side of the river, is the Market-place, containing the Post-office, a few public buildings, one of the thermal springs, a statue of Charles IV., and a monument in memory of the plague, which at the beginning of the last century devastated Bohemia, leaving Carlsbad, however, untouched. From this, a wide street leads up the hill to the Schlossbrunnen. Beyond the market place on the water side is a colonnade and raised platform, near which four of the principal springs rise, and a little lower, opposite to the Hotel de Russie, is the new bath-house, and the large military Sanatorium, erected in 1855.

There are few public amusements of any kind, a small theatre excepted, in Carlsbad. But this Spa enjoys the advantage over some of the gayer watering places on the Rhine, that public gambling is not allowed nor practised here, and moreover there is a very beautiful and interesting country on every side around Carlsbad to induce the visitor to take open air exercise. There are plenty of good hotels, but they are all expensive. In fine, Carlsbad offers no attraction except for those who really stand in need of its very active medicinal waters.

The altitude renders the climate of Carlsbad remarkably changeable, and even in summer, the morning air is often bitterly cold, especially when the north or east winds, to which this valley is exposed, predominate.

The town is built immediately over a vast subterranean boiler, covered in by deposits of the salts contained in the water, which evaporating has left them behind, and on the roof, or crust, thus formed, the town stands. The extent of this abyss, or cavern of boiling water is quite unknown, as all attempts to fathom its depth have failed hitherto. The crust, however, which intervenes between the town and this cauldron, is nowhere thicker than three feet, and in some places appears to be hardly as many inches. It therefore, I think, requires no great sagacity to predict, that in the event of any earthquake, the inhabitants of Carlsbad would very probably try a longer bath in the Sprudel than any of its physicians would prescribe.

Carlsbad is too remote, dull, and expensive, to attract, like the rival Spas of the Rhine, any but real invalids and their attendants. Thus, last year, as I ascertained by the *Kur-liste*, from the 1st of January to the 27th of April, the total number of visitors to Carlsbad was thirty-seven. The season now commenced, and from the 27th of April to the 21st of September, no less than ten thousand four hundred and twenty strangers entered the town. Of these, comparatively few were English; thus on the 28th of June, one hundred and eighty-eight arrivals were recorded, of which only seven were English. The great majority on the list were Russians, next in frequency were Poles, Prussians, Austrians, Italians; the relative numbers of the other nationalities that visited Carlsbad, most of whom were German, I did not ascertain. By the middle of September the season is almost over here, and on the 27th of that month seven thousand one hundred and ninety-one visitors had already left.

The first morning I went out early to visit the springs I was greatly surprised to see the crowd of people in the usually deserted-looking streets which lead to the wells. Most of these, ladies as well as gentlemen, were dressed in a very *negligé* morning costume. Every man, woman, and child, carried a large beaker of steaming water in their hand, and were all gravely engaged in sucking the warm fluid through a small glass tube. And although this may be a very pleasant method of imbibing an iced sherry-cobbler on a hot summer's day, it certainly did not strike me as an agreeable mode of swallowing a large tumbler full of tepid Glauber salts and water at six o'clock on a damp autumn morning.

There are no less than nine thermal springs in Carlsbad which are used medicinally. Although these springs all rise from the same source, yet they differ materially in taste and temperature, according to the distance at which they issue from their common origin, and the nature of the strata they subsequently percolate. But still in composition they so closely resemble the Sprudel, that the analysis of that spring will serve for all the others, and any difference in this respect will be pointed out in the subjoined detailed account of each source. According to Mr. H. Göttl's analysis in 1856, the following table shows the—

COMPOSITION OF THE SPRUDEL.

| Contents of one pint of water. | Grains. |
|---|---|
| Sulphate of soda ... | 14·9606 |
| Sulphate of potash ... | 9·3696 |
| Chloride of sodium ... | 8·7245 |
| Carbonate of lime ... | 2·0198 |
| Carbonate of magnesia ... | 0·3994 |
| Carbonate of protoxide of iron | 0·0307 |
| Alumina ... | 0·2159 |
| Silica ... | 0·0520 |
| Total solid ingredients ... | 44·8340 grs. |
| Gaseous contents in cubic inches. | |
| Carbonic acid ... | 7·80337* |
| Nitrogen ... | 0·03181 |

Dr. Mannl calculates that 15,944 cwt. of Glauber salt, 13,000 cwt. of carbonate of soda, 10,000 cwt. of common salt, and 2,500 cwt. of carbonate of lime are annually discharged into the river from the Sprudel, and totally lost.

The great spring to which Carlsbad owes its fame, as a watering place, is the Sprudel, which is situated almost exactly in the centre of the town, on the bank of the river, at the extremity of a covered colonnade two hundred feet long. The water issues under a kind

* By Berzelius and some other chemists, the amount of carbonic acid gas contained in a pint of the Sprudel is estimated at twelve cubic inches.

of open turret, through which the constantly ascending cloud of steam that escapes is visible for miles around. Under this a large iron fountain is placed, containing a second basin within it, from which the Sprudel issues in remarkable jets, resembling on a vast scale those that escape from a divided arterial trunk, but intermitting their pulsations however, and varying in volume and height from eight inches to as many feet, the average height of the jets being about three or four feet. I, however, observed that on some days the action of the water was not by any means as forcible as on others. The metal basins were thickly incrusted with the white calcareous incrustations of "Sprudel-stone" deposited by the water, which issues, as I ascertained with some difficulty, at the temperature of one hundred and seventy degrees.

The *Hygieas-Quelle* source is found in a small chamber to the right of the Sprudel, from which it is only separated by a passage. This spring certainly owes little to any artificial embellishment, issuing by a tin spout from a wooden trough at one end of the room. The temperature of the water is about five degrees higher than the Sprudel. Its taste, though slightly alkaline, is agreeable. It seems little resorted to, and I seldom met more than three or four votaries at any time round this shrine of Hygeia.

The Marketbrunnen is one of the best attended springs, and seems to be a special favourite with the fair sex. It is situated in a dark room, behind a little pavilion, in a street called after it. The temperature was one hundred and thirty degrees, and the taste is salty and somewhat acidulous.

The Mühlbrunnen rises opposite the bridge, at the extremity of a long colonnade, on the left bank of the river, from a small circular basin, into which two thin thread-like spouts of water, at the temperature of one hundred and thirty-two degrees, discharge themselves. A little farther down the same colonnade is the Neubrunnen; the water is of a deeper colour than the other springs, and more saline in taste; its temperature is a hundred and forty-two degrees.

The Bernardsbrunnen lies in a kind of cellar, under the covered platform, which extends along the left bank of the river, opposite to the Hotel de Russie. The temperature is one hundred and fifty-four degrees, and the sides of the well are thickly incrusted with the calcareous "Sprudel-stone." Behind the platform I have alluded to, in a garden, or terrace, cut out of the side of the mountain, is the Theresienbrunnen, which, next to the Sprudel, is the most celebrated of the springs of Carlsbad, and is now more resorted to by fashionable invalids than the latter; its action, though equally aperient, being said to be much less exciting than the Sprudel. This source is enclosed within a light iron pavilion, and issues at a temperature of one hundred and thirty degrees. The taste is very peculiar, and difficult to describe. When first tasted it seems a combination of saline, bitter, acidulous, and slightly chalybeate flavours, all at once; if it be allowed to remain in the mouth for a moment it now changes completely, and becomes somewhat sweetish, and extremely like what my companion (Mr. Egan) compared it to, i.e., the sky-blue, or weak milk and water, of the boarding-schools. Whether it be this peculiar taste of that which Lord Byron has apostrophized as the type of " Youth and Innocence," which endears it to the fairer portion of creation or not, I cannot say, but certainly this spring is the great centre of attraction for ladies. When I first visited it at half-past six o'clock in the morning the fair valetudinarians were walking up and down in great numbers, sipping beakers of this fluid, to the accompaniment of a tolerably good band. Half-a-dozen white-coated Austrian officers, who were most unremitting in their attentions to the ladies, most of whom were, however, of a certain age, and not of the most attractive appearance in their morning deshabille; together with a few Russians, two or three long-haired, dirty looking Saxon students, and a number of Israelites, from Frankfort, made up the assembly. I should not omit to add that one corner of the promenade was, by common consent, left free to a gentleman in black, with a thick moustache, who held brief colloquies in turn with most of the people present, and who was evidently a brother professor of the healing art.

The other springs which we examined were the Spitalsbrunnen, which is not used internally; the Felsenquelle, temperature one hundred and twenty-nine degrees; and the Schlossbrunnen, temperature one hundred and twenty-nine degrees. This last spring contains more carbonic acid gas than any of the others.

None of the Spas of Europe have been more assiduously "written up" than Carlsbad; but none of the long list of doctors who have extolled the efficacy of these waters, and their own skill, since Dr. Payer, of Elbogen, first chronicled their virtues in 1522, have achieved such celebrity as the Bohemian poet Lobkowitz, whose elegant Latin ode to the Sprudel has been translated into almost every European language. Lord Alvanley and the late Dr. James Johnston have both rendered this ode into English; and a few lines from the latter version will suffice to show the spirit of the original,—

> "Sacred Font! flow on for ever,
> Health on mankind still bestow;—
> If a virgin woo thee—give her
> Rosy cheeks and beauty's glow;—
> If an old man—make him stronger,—
> Suffering mortals soothe and save,
> Happier send them home, and younger,
> All who quaff thy fervid wave!"

Like most other mineral waters, Carlsbad has been extolled as the panacea for almost every disease in the interminable catalogue of human infirmities. One recent writer (Dr. Porges)* enumerated no less than one hundred and sixty-nine maladies curable by Carlsbad water, some of which he sub-divides into six or seven minor ailments. He commences with the letter A.,

\* *The Mineral Waters of Carlsbad, from a Homœopathic point of view.* By Dr G. Porges. Prague, 1864.

and works his way through the alphabet as far as W. In this list the spleen comes in for no less than eight distinct maladies, most of which we are utterly ignorant of in this country, but have the satisfaction of learning that they may all be cured by sending the patient to Carlsbad. We might, indeed, send the majority of our clients there, if the result could be ensured which Dr. Porges tells us occurred to one of his patients, "who got a most flourishing look; nay, even his mental faculties were highly raised in consequence of this full recovery."*

The older physicians seem to have prescribed these waters in what we should now regard as very "heroic" doses. Thus Dr. Mannl states that formerly the patients swallowed from twenty to sixty cups of the Sprudel, and were accustomed to remain eight hours, and even longer, at a time in the bath.†

The physiological effect of the ordinary dose of the Carlsbad water is first an agreeable sense of warmth in the stomach, then a slight perspiration and copious diuresis. The aperient effect depends much on the constitution of the patients, four or five cups being, however, in general sufficient to occasion this. The appetite is sharpened for a few days, but then begins to decrease, and a positive aversion to food often succeeds. A sense of weakness and lassitude is frequently produced at first, but this wears away, and the

---

\* Dr. Porges, *op cit.*, p. 137.
† Dr. Mannl—*Carlsbad and its Mineral Springs*, p. 15. Carlsbad, 1858.

patient gradually becomes stronger and more energetic than he was.

The pathological effect that may be produced by the Carlsbad water is sufficiently important to demand some notice. A tendency to hæmorrhagic and congestive affections has been frequently observed here among patients, and ascribed to the mineral water. Vertigo, spots floating in the air before the eyes, and buzzing in the ears, are symptoms which should by no means be neglected. I myself experienced a very considerable degree of vascular excitement, slight feverishness, and want of sleep, resulting from the quantity of water I was obliged to take while examining the springs here.

The Carlsbad springs are beneficial in all affections dependant on an obstructed or torpid condition of the abdominal viscera. They are, therefore, especially useful to valetudinarians of an obese habit of body, and, conjoined with low diet and abundant exercise, will reduce corpulency more efficaciously, and far more safely, than the absurd system which, under the fitting auspices of a London undertaker, has recently been so productive of evil in this country.

In some affections dependant on an impoverished state of the blood, the Carlsbad waters act more beneficially than any of the strong ferruginous waters. This is probably owing to their stimulant and deobstruent qualities, by which the constitution is prepared for the action of the very small quantity of iron Carlsbad water contains.

In certain cases of misplaced, retrocedent, and lurking gout, and rheumatic-gout, we have sufficient evidence to prove the occasional efficacy of these thermal sources. There is perhaps no Spa so applicable to the majority of cases of hypochondriasis as that now under consideration. The purgative effect of the water, and its action on the liver (the retained and vitiated excretions from which are so often the cause of hypochondriasis), explains its curative effects in this disease.

The writers on Carlsbad also recommend these waters in cases of dyspepsia of an atonic character, and marked by languor of the digestive functions. In chronic diseases of the liver, whether occasioned by long residence in tropical climates or other causes, if there be no inflammatory tendency present, these springs are often very powerful therapeutic agents. In some forms of jaundice the rapidity of their action is said to be remarkable, and their power of dissolving or eliminating biliary calculi not less so. They act in a similar manner, though not so certainly, on some urinary calculi. Carlsbad has been recommended in cases of dropsy, but I cannot concur in this recommendation, as probably no case of dropsy is unconnected with organic disease.

The Carlsbad waters are contra-indicated, that is, are detrimental, in all serious *organic* diseases; they are especially injurious in pulmonary disease, and more particularly so in all forms and stages of consumption. They should be forbidden when there is any inflammatory action present, and even where there may be danger of calling latent inflammation into active existence. Their use would be improper in cases of hemorrhage, and most dangerous where there is any tendency of blood to the head, or cerebral excitement.

The table d'hotes at Carlsbad differ from those of almost all other German watering places, in the simplicity of the fare and the paucity of the dishes. I have seen a newly arrived guest who allowed a few dishes to pass untasted, reserving himself for imaginary *plats* to come, surprised to find dinner concluded before he had tasted anything but the soup. In truth, invalids have here little temptation to gastronomic excess, and perhaps this explains in some degree the good effect of Carlsbad on dyspeptic and gouty patients.

## CHAPTER XVI.

### MARIENBAD AND FRANZENSBAD.

Journey from Carlsbad to Marienbad—Bohemian Scenery—Character of the Peasantry—MARIENBAD—Coup d'œil of the Town—Number of Visitors—Hotels, Doctors, and other *Agrémens*—Analysis of the various Mineral Sources—Their Action on divers Diseases—The Baths—FRANZENSBAD—Its Topography—The Springs and Mud Baths—Their Composition and Medicinal Effects.

OUR last journey had given us a complete disrelish for night travelling in Bohemia; so having made all our arrangements over night, we left Carlsbad soon after daybreak, on what, at first starting, seemed a chill damp morning, but which, before our carriage had well got clear of the town, changed into a bright, hot, sunny forenoon. Our route at first ran along the river, under a complete arch of trees, for a couple of miles, passing a number of neat-looking country houses, "beer-gardens," and cafès, evidently the summer resorts of the people of Carlsbad. We then traversed a wide-spread pine forest, in which an extensive clearance was being carried on; one of the newly-fallen trees lay right across the path, and we had a delay of some minutes while it was being removed.

Somehow or other there is something melancholy in seeing a fine old forest, such as this, whose trees have braved time and the elements for hundreds of years, suddenly felled and sold off, in all probability to pay the gambling debts of some young sprig of nobility at Vienna or Baden.

The way was, however, at length cleared, and we pursued our road, which for some miles now crossed a wide pasture valley; in the midst of which is the ancient-looking town of Petschau, where we had a short delay, while our driver baited his horses and refreshed himself. Soon after passing Petschau this plain again contracts between high mountains, two of which we crossed, and arrived in Marienbad after a drive of seven hours. The whole labour of the thinly-peopled country we had passed through seemed to be carried on by poorly-clad, sunburnt women, who were engaged ploughing, digging, tending cattle, &c., whilst all the men we met with on the road were pursuing the " dolce far niente," and were comfortably and even tastefully dressed, in a costume not very unlike the Andalusian. Everyone, as we passed, saluted us, and as we of course returned the civility, by the time we got to Marienbad the brims of our hats were considerably the worse for our politeness. We had, too, a proof of the honesty of this people afforded us, it is true in a very small matter, but " ex pede Herculem,"—Mr. Egan dropped a book out of the carriage, and did not discover his loss for some time, when he stopped and went back with the driver to look for it on the road; the search proved useless, and he was just returning to the carriage, when a man ran up to him, and handed him the volume which he found nearly a mile off, and had pursued the carriage with it. Nor was it in the hope of reward, for he refused to accept of any remuneration for his trouble.

MARIENBAD is, I think, one of the prettiest of the continental Spas. The town, which is quite modern, consists chiefly of large hotels and lodging-houses. It stands on the declivity of the Steinhau, a pine-clad hill facing the south, and forms an incomplete square, built around three sides of an extensive park or garden, which contains most of the mineral sources. Nearly in the middle of this park, is a very handsome church, built in the shape of a Greek basilica. This belongs to the monks of Töpl, who are the proprietors of the town.

The hotels are of enormous size, and look quite out of proportion to the little town in which they stand. But they are well filled however, in the summer and autumn months; for I saw by the list that no less than five thousand five hundred invalids had visited Marienbad up to the 30th of September, after which very few valetudinarians arrive, as the season closes early. Where there are patients, there will of course be doctors; and this is the case in Marienbad, where a dozen physicians, besides three surgeons, reside in the various hotels from May to September.

The largest hotel here, and I believe I may add one of the largest in Germany, is Klinger's, next to which is the Sonne. The Hotel de Leipsic, in which we stopped, though nearly as large, was not at the time of our visit —when, however, the season for visitors was over—by any means comfortable; and, besides, was dearer than many of the best hotels in the principal German or Belgian towns.

Behind this square, there is a small theatre, which, however, is not a very imposing temple of Thalia. Marienbad, though one of the most beautifully situated watering places in Europe, must, I think, be a very dull residence for those who do not require its mineral waters. The cursaal, at the extreme end of the park, is a large building, with a magnificent central salon. It was quite deserted however, and two Austrian officers, who were playing billiards, seemed quite delighted at the arrival of a couple of strangers to break the solitude of this immense hall.

Nothing can be prettier than the arrangement of the park. Near the Kreuzbrunnen is a long salon for exercising in wet weather; opposite to this is a little orchestra, under a pavilion, where bands perform twice daily; moreover, the different sources and fountains are all kept in perfect cleanliness, contrasting in this respect with many of the Spas we had visited, where every kind of rubbish is suffered to accumulate in the mineral springs.

The principal sources here are those whose analysis is subjoined :—

COMPOSITION OF MARIENBAD MINERAL SPRINGS.
ACCORDING TO SUTRO.

|  | Kreuz-brunnen. | Karolinen-brunnen. | Ambrosius-brunnen. | Ferdinands-brunnen. |
|---|---|---|---|---|
| Sulphate of soda | 38·11 | 2·79 | 1·86 | 22·53 |
| Chloride of sodium | 13·56 | 0·82 | 1·64 | 8·99 |
| Carbonate of soda | 7·13 | 0·20 | 1·66 | 6·13 |
| Carbonate of lime | 3·93 | 3·66 | 2·89 | 4·01 |
| Carbonate of magnesia | 2·71 | 3·94 | 2·72 | 3·04 |
| Carbonate of iron | 0·17 | 0·44 | 0·34 | 0·39 |
| Carbonate of manganese | 0·03 |  |  | 0· 9 |
| Carbonate of lithia | 0·11 |  |  | 0·06 |
| Silex | 0·38 | 0·46 | 0·48 | 0·66 |
| Total solid contents, in grains | 66· | 14½ | 10½ | 45¾ |
| Carbonic acid gas, in cubic inches | 8½ | 15·43 | 12·9 | 13¾ |

Before entering on the peculiarities of each of these springs, and their special therapeutic effects, it will be necessary to make a few observations on the general effects of the Marienbad waters. From the foregoing table it will be seen that the principal ingredient in the various mineral springs of Marienbad is sulphate of soda, or Glaubers salt. It may be asked, if this be the case why should not a solution of that salt of the same strength answer the purpose, and save the expense and inconvenience of a journey to the "Garden Spa of Bohemia." Having entered so fully into the question of the comparative effects of natural and artificial

*Lectures on the German Mineral Waters. by Dr. Sigismund Sutro,* p. 159. London, 1851.

mineral waters in the introductory portion of this work, I need not discuss that question again, nor repeat my observations on the value of travel and change of living and scene, as adjuncts to mineral waters in the treatment of disease. In the present instance we have, in addition to this, to consider that Marienbad is a compound Spa, so that the remedial effect of the simple sulphate of soda, sold in our apothecaries' shops, which produces thin serous evacuations, and acts as a powerful antiphlogistic, is very different from the gently stimulating tonic and alterative action of a compound fluid, such as Marienbad water. The latter containing nine or ten different salts, in small doses, it is true, but in such minute division and perfect solution that they are more readily absorbed into the system than our grosser preparations.

The Marienbad waters act directly upon the liver, the secretion of which is notably increased by their use. Their first effect is that of a saline aperient, not followed, however, by the debility attending the exhibition of other equally powerful remedies of that class. On the contrary, the carbonate of iron they all contain, though in such small proportions, produces a very decided tonic effect. The appetite is almost invariably sharpened by them; the pulse is generally at first quickened, and the kidneys secrete more copiously under their influence, without the dryness of skin and obstructed perspiration that often accompanies the action of diuretics.

Opposite to the windows of our hotel was the principal mineral spring—the Kreuzbrunnen, which is enclosed in a square chamber, under a dome. In front of this is a very neatly kept garden, surrounded by a piazza with marble pillars, which leads to the pump room. The Spa is forced up by a wheel pump, and issues from a bright silver spout, at the temperature of fifty-five degrees, but before being used is generally mixed with a sufficient quantity of warm water to bring the temperature of the draught up to ninety degrees, and by this admixture its action is said to be quickened, and its taste is certainly improved, now very closely resembling that of the Wiesbaden spring. The usual dose as an aperient is from three to four glasses of this mixture. The special action of the Kreuzbrunnen is to stimulate all the abdominal organs, and especially the liver, to increased action. The anti-spasmodic, alterative, and other qualities which are attributed to the Kreuzbrunnen, are in reality, I believe, only the results of its action on the liver and intestinal canal. In like manner its curative effects in many obscure nervous, hysterical, and uterine affections; as well as in dyspepsia, hypochondriasis, some cutaneous maladies, and enlargements of the abdominal viscere, and, above all, in general and local plethora, are simply owing to its solvent and purgative action. In all these diseases, whatever may be its *modus operandi*, the Kreuzbrunnen is unquestionably a very powerful remedy.

The next source we examined was the Karolinenbrunnen, which is nearly in the centre of the park, immediately behind the church, under a light open pavilion. The well is very deep, and as we could find no one in charge of it we had some difficulty in getting sufficient of the water to examine it, but with a little trouble we effected this. The temperature of this Spa is fifty-five degrees, and its flavour is by no means agreeable, being that of rotten eggs, or sulphurated hydrogen. The sides of the well were thickly incrusted with a yellowish greasy deposit.

This spring is the strongest tonic source in Marienbad, containing about half-a-grain of iron in a tumblerfull of the water, as well as double the amount of carbonic acid gas found in the Kreuzbrunnen. But it contains so little sulphate of soda as to render the use of the Kreuzbrunnen generally necessary conjointly with that of this acidulated chalybeate when an aperient effect is desired. It acts as a stimulant on the vascular system, increasing the frequency of the pulse, and as an excitant on the nervous system. In short, it may be prescribed in most cases of general and local debility requiring a ferruginous tonic.

A few yards to the right of this last spring is the Ambrosiusbrunnen, the taste of which is very piquant, owing to the large amount of carbonic acid gas in it. The effects of this source, which is somewhat alkaline, although rather weaker, are yet essentially similar to those of the Karolinenbrunnen.

The Ferdinandsbrunnen lies about a mile outside the town, and is approached by a very pretty walk through a wood. It is contained in the centre of a long open colonnade. The taste of the water, which is cold, is exactly that of a Seidlitz-powder, and it seems to be a favourite beverage with the inhabitants of Marienbad. This source, as may be seen by the analysis, contains more than twice as much iron as the Kreuzbrunnen, but considerably less of the other salts which characterize that spring, than which its action is consequently less aperient, but more tonic, holding in this respect an intermediate place between the Karolinenbrunnen and Kreuzbrunnen. It is employed in the same class of cases as the latter spring, when a greater amount of debility is present, and in such cases acts more energetically on the torpid intestinal and renal functions.

I have not entered into the analysis of the Marienquelle, which was the original spring, as it is no longer used for drinking. It rises within a primitive-looking wooden building, about eighty feet long, enclosing a large reservoir of the water, which, from the extraordinary amount of gas it contains, seems to be in a constant state of violent ebullition. A gallery runs round two sides of this room just above the level of the stratum of carbonic acid gas, which rises about twenty inches over the water, as I measured by letting down a lighted taper, which on arriving within that distance of the surface was immediately extinguished. The temperature of the water is 52 deg. The taste is bitter and very acidulous ; but it is not used internally, being employed only for baths, to which it is carried by metal pipes from this cistern.

The baths, which are handsomely fitted up, are now regarded merely as adjuncts to the internal use of the waters. The first effect they produce, at blood heat, is a sense of chilliness, which very soon goes off, and is succeeded by a feeling of heat and perceptible redness of the skin. They almost invariably occasion a copious secretion from the kidneys. They are particularly useful in arthritic and rheumatic affections, torpidity of the liver and bowels, old painful ulcers and wounds, loss of muscular power, when not preceded or connected with any cerebral affection, especially of an apoplectic character, in which case their use might very probably be fatal. They are moreover, sometimes prescribed in scrofula, and in some glandular enlargements of a simple nature.

*Mud Baths* are used in Marienbad, but not to the same extent as at Franzensbad, in the following account of which place they will be described.

*Gasbäder*, or gas baths, are also employed here, both locally and generally. In the latter form of bath the patient enters a square box, shaped exactly on the model of a Chinese pillory, covering the entire body except the head, which protrudes through a hole in the lid. Into this he is put, dressed in his ordinary habiliments; the gas is let in through a pipe in the bottom of the box, and presently, as the gas rushes in, a sensation of tingling and pricking is felt creeping up the legs, and gradually extends over the entire body. The " gas-bade" owes its medicinal application to Dr. Sturve, of Dresden, who first employed it on himself when suffering from a paralytic affection connected with disease of the sciatic nerve, and derived extraordinary benefit thereby. Owing to his panegyric on its effects, published shortly after, it rose into great vogue with German physicians and patients in the treatment of some diseases marked by general torpitude and vascular languor, suppressed menstrual and hœmorrhoidal discharges, scrofulous ulcers, and swellings, and other similar complaints.

From Marienbad, a drive of twenty miles, passing through the historic town of Eger, brings us to Franzensbad, which lies about thirty miles south-east of Hof, and nearly forty miles south-west of Carlsbad.

FRANZENSBAD is situated in the midst of a bog, surrounded by bleak-looking mountains, and consists of four streets crossing each other at right angles, built on piles driven into the soft mud beneath. The principal street is the Kaiser-Strasse, a long boulevard planted with chestnut trees. This contains the chief hotels and bath houses, and leads up to the Franzensquelle source, to which the town owes its name.

There are four mineral springs here, the composition of which will be seen by a glance at the following table.

ANALYSIS OF FRANZENSBAD MINERAL WATERS.

|  | Franzensquelle (by Berzelius). | Salzquelle (Berzelius). | Wiesenquelle (Zembsch). | Louisenquelle (Trommsdorf). |
|---|---|---|---|---|
| Sulphate of soda | 24·50 | 21·52 | 25·65 | 21·41 |
| Chloride of sodium | 9·23 | 8·76 | 9·32 | 6·76 |
| Carbonate of soda | 5·18 | 5·20 | 8·97 | 5·49 |
| Carbonate of lithia | 0·03 | 0·02 | 0·02 | — |
| Carbonate of magnesia | 0·67 | 0·79 | 0·61 | — |
| Carbonate of lime | 1·89 | 1·41 | 1·37 | 1·60 |
| Carbonate of protoxide of iron | 0·23 | 0·07 | 0·13 | 0·32 |
| Carbonate of protoxide of manganese | 0·04 | 0·01 | 0·02 | — |
| Phosphate of lime | 0·02 | 0·02 | 0·02 | — |
| Other salts | 0·47 | 0·49 | 0·47 | 0·22 |
| Total solid contents in grains | 42·18 | 38·29 | 46·58 | 35·80 |
| Carbonic acid gas | 40 | 26·88 | 30·69 | 32·53 |

The Franzensquelle was formerly known as the Eger Spring, and its water is still exported under that name. It is situated in a circular colonnaded pump room at the top of the Kaiser-strasse. Its temperature is about 51 deg., and its taste is agreeable and piquant, though there is an after taste of sulphuretted hydrogen. In composition, it resembles the Kreuzbrunnen of Marienbad, containing, however, twice as much carbonic acid gas, though weaker in saline ingredients.

A few yards to the eastward of the last spring rises the Salzquelle, under a long piazza. The taste of this source is slightly alkaline, but not disagreeable. The other sources lie close to each other, behind the the same colonnade.

In their action, the Franzensbad waters resemble those of Marienbad, but, containing less sulphate of soda and more carbonic acid gas and iron, are less lowering, although probably fully as active. They are, moreover, in some cases, very stimulating, and act as excellent tonics in certain anæmic affections connected with gastric and intestinal derangement. The dose varies with the state of the patient and the spring which is employed, on which subject, though a table such as the foregoing may be of aid to a physician at a distance from the Spa, the patient should not rely altogether upon it, but before commencing the course, should consult a local practitioner.

The Franzensbad springs are said to produce a sedative action on the nervous system, while imparting strength and tone to the muscles; they also purify the blood by their purgative and diuretic action, and improve its composition by the additional nutriment which the patient is now enabled to digest. Therefore Franzensbad is largely resorted to by dyspeptic, and, above all, by hypochondriacal patients. It was by the use of this remedy that the gloomy savage, whom Mr. Carlyle has recently held up as the model of all virtue —Frederick the Great—was aroused from one of his fits of hypochrondriacal insanity in 1748.

Near this spring is the Gasquelle, or gas source,

over which baths have been erected. Its effects are precisely the same as those of the gas-baths of Marienbad, and need not be again described.

The Mud-baths are the special advantages of Franzensbad, and are extolled as a cure for every disease under the sun, *et quibusdam aliis*. The soft boggy earth which surrounds Franzensbad on all sides is the material of these baths. It is dug up and repeatedly forced through sieves, until it is perfectly free from all foreign matters, woody fibres, &c., and when it has attained a perfectly soft, homogeneous, condition, it is diluted into a semi-fluid, black, pultaceous mass, exhaling a strongly sulphurous smell, with the Louisenquelle water, and is heated to about 100 deg. Into this uninviting looking bath the patient enters, and so dense is it, that it is generally some time before he can immerse his whole body. The sensations it produces are described as being particularly agreeable, and the patient leaves with reluctance at the expiration of the quarter of an hour, which is the usual duration of the mud-bath. As soon as he leaves this, the patient is placed in a plain tepid water bath, where he finds sufficient occupation for half-an-hour in restoring himself to something like cleanliness.

The principal saline contents of this mud are sulphate of soda, lime, magnesia, iron, and alumina, silica, tannin, sand, resinous and vegetable matters.

The primary action of these baths is stimulant and exciting to the nervous system. They produce some degree of cutaneous irritation ; while in them the skin looks corrugated and wrinkled, but feels smooth and glossy immediately after immersion. The appetite is almost always increased by the external use of this black mineralized mud.

The *Bains de Boue* are used in chronic arthritic and rheumatic affections, in some skin diseases of an obstinate and languid character, and in similar ulcers ; in glandular swellings, in paralytic complaints, particularly of the lower extremities ; and are renowned for the cure of old and painful wounds.

## CHAPTER XVII.

### TEPLITZ AND THE MINOR BOHEMIAN SPAS.

The town of TEPLITZ and its environs—How to arrive there—The mineral waters—Their analysis and action—SCHÖNAU and its thermal sources—BILIN, "the German Vichy"—Diseases in which resorted to—The plain of the bitter waters—PULLNA—The village and springs—SEDLITZ—The wells—Analysis of, compared to our *Seidlitz* powders—SAIDSCHUTZ—Observations on the therapeutic influence of the Bitter waters.

TEPLITZ, which, next to Carlsbad, is the most important of the Bohemian Spas, may be now reached with equal facility from Northern and Southern Germany, being only sixty-three miles from Dresden, sixty from Prague, and three hundred and thirty miles from Vienna, with all of which it is connected by railway.

The town lies in a long, narrow, and very fertile valley, protected by the Erzgebirge mountains, between which and the Mittelgebirge hills this watering-place is situated, and thus enjoys a peculiarly genial climate.

The environs of Teplitz are not less celebrated for the beauty of the scenery than for the fertility of the soil. From the heights above the town, or from the Schlossberg hill, the view embraces a vast and varied perspective, extending from the dark forests and rocky precipices of the Erzgebirge mountains, across to the distant plain of Culm, on the north, which no Frenchman looks at without bitter remembrances of his ill-fated countrymen who were there cut to pieces in 1813, while to the south may be traced the winding outline of the Elbe.

The town of Teplitz, which consists of a few streets of lodging-houses and hotels, most of which are situated in the Lange-strasse, the Mühl-strasse, on the road to Schonau, and the two squares of the market place and Schloss Platz, contains little else but its mineral springs to attract attention, being in all other respects a quiet little country town of about three thousand inhabitants. The Bade Platz, containing the château, which forms one side of it, is completed as a square by the bath houses and hotels. The château and its garden, with the beautifully wooded park behind, serve all the purposes of a "Kur-haus," and contains a small theatre, restaurant, reading and assembly rooms, and is the great rendezvous of all the visitors to Teplitz.

In the immediate vicinity of the town, and in the adjoining village of Schönau, which is really the northern suburb of Teplitz, there are no less than seventeen mineral and thermal springs, differing only in temperature. The hottest of these sources is the Hauptquelle, the heat of which is one hundred and twenty degrees; and the coolest is the Gartenquelle, the temperature of which is eighty degrees, and which is the only source used internally. These springs are all saline, alkaline, and slightly chalybeate, but are none of them of any chemical strength sufficient to explain their undoubtedly powerful medicinal effects, which in great measure must be accounted for by the high temperature at which they are employed. The solid ingredients are about five grains to the pint of water, and chiefly consist of carbonate of soda.

Subjoined is the analysis of the principal spring of Teplitz, as given by Dr. Sutro,* whose table differs somewhat from M. Wolf's, as cited in Dr. Seegen's work on mineral waters† :—

COMPOSITION OF THE HAUPTQUELLE.

| | |
|---|---|
| Sulphate of potash | 0·43 grains. |
| Carbonate of soda | 2·68 ,, |
| Carbonate of lithia | 0·01 ,, |
| Carbonate of lime | 0·32 ,, |
| Carbonate of strontia | 0·01 ,, |
| Carbonate of manganese | 0·08 ,, |
| Carbonate of magnesia | 0·05 ,, |
| Carbonate of iron | 0·03 ,, |
| Chloride of sodium | 0·43 ,, |
| Chloride of potassium | 0·10 ,, |
| Iodide of potassium | 0·05 ,, |
| Phosphate of alumina | 0·02 ,, |
| Silico-fluoride of sodium | 0·13 ,, |
| Silica | 0·31 ,, |
| Crenic acid | 0·09 ,, |
| Total solid ingredients in sixteen ounces of water | 4·84 (or rather less than 5 grs.) |

It would be of no practical utility to describe the

* *Lectures on the German Mineral Waters*, by Sigismund Sutro, M.D., p. 25.
† *Handbuch der Heilquellenlehre*, von. Dr. Josef Seegen, p. 663. Wien, 1862.

seventeen thermal springs of Teplitz seriatim, as the only real difference between them is in temperature. But yet they differ so materially in their action, notwithstanding the identity of their chemical composition, that it would be quite unsafe to use any of them without first consulting a local practitioner as to the proper spring to be used in each case. The general action of the Teplitz baths is to produce a considerable degree of vascular excitement, slight feverish disturbance, and after some days they occasion a reddish cutaneous eruption, with great irritation of the skin. If their use be still persisted in, they give rise to the symptoms I have already elsewhere described under the name of "Bad-sturm," or saturation fever.

Such are the effects of the hotter springs of Teplitz, and therefore many practitioners enjoin their patients, before trying them, to commence with the cooler waters of the neighbouring village of Schönau.

These springs are chiefly employed for bathing, with the exception of the Gartenquelle, the internal use of which is, in some cases, conjoined with a course of the baths. Its action, thus used, is that of a mild aperient and resolvent; it is also said to have a wonderful power of causing the absorption of chronic glandular and visceral enlargements.

Used as baths, the Teplitz waters are employed, and as the resident physicians assert, with remarkable success in the treatment of certain forms of paralysis, though

I think their utility in such cases must be exceedingly limited. They are used also in chronic rheumatic arthritis and diseases of the joints; in spinal curvature and hip-joint disease; in enlargement of the liver and spleen. The baths are moreover resorted to in some affections peculiar to women: suppression of habitual discharges, chronic skin diseases, hysteria, hypochondriasis, and other maladies, in which a remedy which combines stimulant with resolvent and tonic properties is required.

I need hardly add that mineral springs possessing these qualities are especially contra-indicated, in all cases which, as I have shown in the preceding chapters, are unsuitable for the use of any warm baths,—namely, plethoric, inflammatory, or hæmorrhagic diseases.

Within nine miles of Teplitz, on the road to Prague, is the small town of Bilin, the so-called "Vichy of Germany." This little market town of three thousand inhabitants lies about a mile from the remarkable basaltic mountain of the Biliner-Stein, and needs no description here, being seldom resorted to by invalids, although its waters are largely exported.

The chief ingredient of the mineral water of Bilin is carbonate of soda, of which it contains twenty-three grains in the pint, being five grains stronger than the Fachingen water, to which it otherwise bears a close resemblance. There are four or five mineral springs here, the composition of which are almost identical. The chief of these sources is the Josefsquelle, the analysis of which, according to Dr. Redtenbacher, is as follows :—

COMPOSITION OF BILIN MINERAL WATER.

| | |
|---|---|
| Carbonate of Soda | 23·106 grains |
| Carbonate of Lime | 3·089 ,, |
| Carbonate of Magnesia | 1·098 ,, |
| Carbonate of Protoxide of Iron | 0·080 ,, |
| Carbonate of Lithia | 0·110 ,, |
| Sulphate of Potash | 0·985 ,, |
| Sulphate of Soda | 6·350 ,, |
| Chloride of Soda | 2·935 ,, |
| Basic Phosphate of Alumina | 0·065 ,, |
| Silex | 0·244 ,, |
| Total Solid Ingredients | 38·062 ,, |

The Bilin waters are employed in cases requiring an alkaline carbonated water, in diseases of the urinary organs and kidneys, in Bright's disease, in certain cases of jaundice, and in some forms of arthritic and rheumatic affections of the joints.

On the route between Teplitz and Carlsbad, and within some twelve miles of the former town, are three of the most remarkable mineral springs in Europe. As they are not watering places, however, a very brief account will here suffice of the "bitter waters" of Püllna, Saidschütz, and Sedlitz.

The village of Püllna, on the road to Carlsbad, is the chief source whence the "Bitter Wasser" is exported to every part of the civilized globe; but as the water is very seldom drank on the spot, there is hardly any accommodation for invalid residents, beyond a very second-rate inn. The mode of collecting the "Bitter-wasser" at Püllna is similar to that employed at Sedlitz, and Saidschütz, and may here be described once for all. These three sources are situated in an extensive marly plain, the soil of which for a limited area round each spring has a peculiar light yellowish colour, and is perfectly destitute of vegetation. In this marly clay, wells and tanks are dug, and the rainfall and oozing of the soil is suffered to accumulate in them for months, dissolving out the soluble saline ingredients from the subjacent formations.

To return to Püllna—according to Professor Ticinus the following is the analysis of

### PÜLLNA BITTER WATER.

| | |
|---|---|
| Sulphate of magnesia | 96·975 grains. |
| Sulphate of potash | 82·720 ,, |
| Sulphate of soda | 10·125 ,, |
| Chloride of magnesium | 19·120 ,, |
| Nitrate of magnesia | 4·602 ,, |
| Crenate of magnesia | 4·640 ,, |
| Carbonate of magnesia | 2·280 ,, |
| Sulphate of lime | 0·800 ,, |
| Carbonate of lime | 0·760 ,, |
| Bromide of magnesium | 0·588 ,, |
| Phosphate of soda | 0·290 ,, |
| Total solid ingredients | 222·900 ,, |

Together with 49 cubic inches of carbonic acid gas; in sixteen ounces of the water.

To the eastward of the last described source in the same plain lies SEDLITZ, or Seidlitz, whose name is the most familiar, and whose waters are the least used of all the German mineral springs. Sedlitz is a wretched-looking place, hardly meriting the name of a village, and the wells whence the water should be derived are a few shallow, circular pits, whose contents very seldom find their way to this country. The actual Sedlitz water differs in every respect from the "Genuine Sedlitz powder" of our chemists. Instead of the cooling, agreeable draught used in England under that name, the true Sedlitz-Wasser is a yellowish, somewhat oily-looking fluid, with a nauseous, intensely bitter taste. Its composition is not less different from its English namesake than its taste, the following being the ingredients contained in sixteen ounces of the—

SEDLITZ WATER.

| | |
|---|---|
| Sulphate of magnesia | 104 grains. |
| Sulphate of lime | 8 ,, |
| Carbonate of lime | 8 ,, |
| Chloride of soda | 3 ,, |
| Carbonate of magnesia | 3 ,, |
| Total solid contents | 126 ,, |
| With 3½ grains of carbonic acid gas. | |

Whereas a *genuine (English) Sedlitz powder* consists of Rochelle salt, or tartarate of soda and potash, and bicarbonate of soda, set into effervescence by half a drachm of tartaric acid. Thus, as may be seen by a glance at the foregoing table, our "Sedlitz powders" do not contain a grain of any of the ingredients of the Sedlitz Bitter-wasser.

Half-an-hour's walk to the eastward of Sedlitz, on a slight elevation above the plain is SAIDSCHUTZ, the most largely exported of these waters. It is considerably stronger than the Sedlitz, and is more varied in composition. Thus, according to Berzelius, the following is the analysis of sixteen ounces of

### SAIDSCHÜTZ WATER.

| | |
|---|---|
| Sulphate of magnesia | 84·16 grains. |
| Nitrate of magnesia | 25·17 ,, |
| Carbonate of magnesia | 4·98 ,, |
| Chrenate of magnesia | 1·06 ,, |
| Chloride of magnesium | 2·16 ,, |
| Sulphate of potash | 4·09 ,, |
| Sulphate of soda | 46·80 ,, |
| Sulphate of lime | 10·07 ,, |
| Oxides of manganese, iron, tin and copper, &c. | 0·28 ,, |
| Total solid constituents | 178·77 ,, |

These "Bitter-Waters," though differing in strength, and also, to a small extent, in composition, all resemble each other sufficiently to enable us to describe their action together. They are all aperient, resolvent, and diuretic, varying in strength from the Püllna, which is the most powerful, to the Sedlitz, which is the weakest. They are all too strong, in general, to be used undiluted, and their action is quickened as well as rendered safer by mixture with an equal amount of warm water. They closely resemble the Friedrichshall bitter-water, and are employed in the same class of cases—namely, in habitual torpidity of the intestinal canal, in congestions, torpidity, and enlargement of the liver and spleen, in plethora, in tendency to apoplexy and congestion of the brain and lungs, and similar diseases in which depletion is indicated. I have prescribed them with advantage in some forms of dyspepsia, and I have no doubt that many cases of hypochondriasis would be served by a course of these bitter-waters. The usual dose is half a bottle of the water with an equal quantity of tepid water every morning, which may be repeated within a couple of hours, if necessary.

## CHAPTER XVIII.

### CANNSTADT AND WILDBAD.

CANNSTADT, its Position, Population, and Environs—A German Festival and our Visit—The Mineral Springs—Their Composition and Effects—Class of Patients by which they are used—Journey to Wildbad—The Black Forest and Valley of the Enz—A troublesome Companion—Arrival in WILDBAD—Hotel Klump—Dr. Haussman—Description of this Spa—Detailed Account of the Bathing System—The Hospital—Curious Privilege—Geological Formation—Recent Analysis of the Water—Its Mode of Action explained—Newly-discovered Sources—Physiological Effects of the Baths—Spa Fever—Diseases in which Wildbad is resorted to.

WURTEMBERG, though not particularly rich in mineral waters, yet contains two of these fountains of Hygeia, namely Cannstadt and Wildbad, both of which, and especially the latter, are of great therapeutic importance. The first of these, Cannstadt, is a small town of some four thousand inhabitants, on the right bank of the Neckar, about four miles from Stuttgardt, with which a continuous park connects it. The environs, which are extremely picturesque and fertile, abound in corn and vine, and form what is termed "The Garden of Suabia." At the time of our visit, the town presented a lively and animated aspect, owing to the expected arrival of the newly-crowned King of Würtemberg, who was that day to make his first public appearance, since his accession to the throne, at the annual races and fair of Cannstadt. Triumphal arches were being constructed; banners hung from every house, and trees, temporarily transplanted, suddenly converted the streets for the occasion into regular lines of boulevards. Under ordinary circumstances, however, Cannstadt, despite the beauty of its situation, must, I think, be a very dull little town. The hotels, though tolerably good, seem to do but little business, as the great majority of the valetudinarians who drink these waters are citizens of Stuttgardt, who come by train for that purpose, and then return to the capital.

There are eighteen or twenty mineral springs in Cannstadt, which are said to differ from each other considerably, and some writers enumerate nearly double that number, and attribute specific properties to each; this is, however, I think, perfectly unnecessary. For, as far as I could ascertain, though certainly differing from each other in some respects, these mineral springs all belong to the class of saline chalybeates, and originate in a strata of ferruginous limestone covered by slate and marl.

## ANALYSIS OF THE PRINCIPAL SPRINGS OF CANNSTADT (FEHLING).

|  | Sulzerrain-quelle. | Frösnerische-quelle. | Sprudel. | Neuquelle, No. 1. | Neuquelle, No. 2. |
|---|---|---|---|---|---|
| Chloride of Sodium | 16·29 | 19·50 | 16·42 | 12·63 | 7·59 |
| Chloride of Potassium | ... | 0·25 | ... | 0·87 | 0·57 |
| Chloride of Magnesium | ... | 0·18 | ... | ... | ... |
| Carbonate of Lime | 7·89 | 7·38 | 8·82 | 7·95 | 6·40 |
| Carbonate of Magnesia | ... | 0·31 | ... | ... | ... |
| Carbonate of Protoxide of Iron | 0·16 | 0·25 | 0·18 | 0·17 | 0·02 |
| Sulphate of Soda | 2·92 | 4·75 | 2·18 | 0·87 | 1·04 |
| Sulphate of Magnesia | 3·53 | 2·25 | 3·51 | 3·89 | 3·34 |
| Sulphate of Lime | 6·43 | 7·75 | 6·32 | 6·88 | 5·06 |
| Sulphate of Potash | 1·23 | ... | 1·38 | ... | ... |
| All other Ingredients | 0·16 | ... | 0·17 | 0·09 | 0·08 |
| Total Solid Contents | 38·61 | 42·62 | 38·98 | 33·45 | 24·10 |
| Carbonic Acid Gas | 23·5 | 19·4 | 27·7 | 14·6 | 8·8 |

The Cannstadt springs owe their efficacy to the combination of different purgative salts, together with a small quantity of lime, rendered peculiarly active and soluble by an excess of carbonic acid gas. Thus they combine aperient with slightly tonic properties. They are, therefore, valuable deobstruent remedies, and are particularly useful in glandular obstructions occasioned by want of power in the system. They are also frequently prescribed by the physicians of Stuttgardt in dyspeptic cases, and under their use the digestive organs not only regain their tone, but are moreover, said to acquire a positive increase of power.

The principal medical use of the Cannstadt springs

is however, in cases in which a mild tonic and solvent is indicated, in chronic catarrhal affections of the mucous membranes, and in some forms of scrofulous disease.

The "Sprudel" spring is that most frequently prescribed, and in its mode of origin somewhat, though indeed on a very small scale, resembles its namesake of Carlsbad. The temperature of the water is about sixty-five degrees, and its taste is slightly ferruginous, piquant, and agreeable. There is another celebrated source which is used principally for bathing, "Die obere Sulz," a small pond of about half-an-acre diameter, formed by several springs. The water is so gaseous that it seems absolutely boiling, so hissing and bubbling is it with nitrogen and carbonic acid gases. The temperature is about sixty-six degrees, and it contains a quarter of a grain of carbonate of the protoxide of iron to the pint of water. Its medicinal action is chalybeate and solvent, and it is employed externally, in the form of ascending douche, in the treatment of catarrhal affections of the mucous membranes.

WILDBAD, the most romantically situated of the German watering places, is beautifully placed in the very centre of the Black Forest, in the long, narrow valley of the Enz, about six leagues from Pforzheim, the nearest railway station.

Nothing can excel the quiet beauty of the road from Pforzheim through the Black Forest to Wildbad.

During this drive even the most unobservant traveller must be struck by the evidences he may here see of the patient industry of the people of this country, shown by the labour and agricultural skill with which every available inch of ground between the river and the forest is cultivated. This is only a narrow strip on each bank of the Enz, but however, it is tilled with a care which throws Mr. Mechi's model farming completely into the shade. Through the midst of this valley the Enz, the noisiest and most turbulent mountain torrent of its size that can be imagined, rushes, white with foam, pefrorming miniature waterfalls and cataracts every half-mile or so, bearing rafts so narrow, although longer than the Great Eastern, that they hardly afford footing to the man who navigates them, down to the distant Rhine; while back from the river extend for miles on either side the gloomy pine shades which are so well characterised as the "Black Forest."

Our enjoyment of this drive was, however, considerably marred by the intrusive impertinence of a small, rotund, over-dressed, bejewelled, talkative German; who told us that he was a native of Wildbad, returning to see his friends, after an absence of sixteen or eighteen years in England and the United States, during which time he had made his fortune, and certainly seemed to have half forgotten his own tongue without acquiring ours. One thing he had at least succeeded in picking up to perfection, this was the inquisitiveness peculiar to Yankee travellers; and having, unasked, told all about himself, he imagined that he had a right to exact the same information about us. He asked where we were going? why were we going? where did we come from? what was our business, fortune, name, age, place of abode, and family? he next offered to exchange watches, which, as his was the most brilliant of pinchbeck, and mine happened to be a rather valuable timepiece, left me as a legacy by an old patient, would not have been such a bad speculation for him. On our declining either to enter into conversation or trade with him, he expressed his indignation so strongly that we were at last obliged to intimate, in the clearest way, our opinion that he would find the outside of the coach more comfortable, in which arrangement, as we had secured the three inside places beforehand, he was induced to acquiesce.

At last we arrived in Wildbad, and descending at the Hotel Klumpp, which, I may remark, is the chief hotel in Wildbad, and is one of the most comfortable and cleanest hotels in Germany, we engaged our rooms, ordered dinner, and went out to present our letter of introduction to Dr. Haussmann, the principal physician of the Spa. Dr. Haussmann, who received us with great courtesy, afterwards accompanied us to every object of interest about Wildbad, and I am largely indebted to him for the information I succeeded in gathering concerning this watering place.

Wildbad, which stands in the narrowest part of the valley of the Enz, contains a population of about two thousand five hundred inhabitants, and, excepting the bath house and hotels, has hardly a single building of any size. The church, which is a plain, unpretending edifice near the springs, affords a proof of German "liberalism" in religious matters, being each Sunday used in succession by the Catholic and Protestant congregations. The Würtemberg Government have already expended about eighty thousand pounds on the improvement of Wildbad, and contemplate further improvements. There is some, but, I trust unfounded fear entertained lest the curse of a public gambling table may be introduced into Wildbad.

The Grand Bath House, which is almost *vis-a-vis* with the Hotel Klumpp, is the most perfect bath establishment in Europe for its size. Of its extent some idea may be formed from the fact that twelve hundred baths can be daily administered, each bath being moreover, of considerable duration. The total number of baths given last season amounted to upwards of eighty thousand. The foundation of this building is cut out of the solid granite rock, through which the water percolates, and on which the baths lie, with the intervention of a thin layer of fine sand. The establishment is equally apportioned to male and female baths. In both are large public "piscina," and smaller cabinets for those who prefer bathing separately. One characteristic of all these baths is the great height of the rooms, so that one is not plunged into a hot vapour bath before entering the water, as is the case in almost every other bathing establishment in Germany that I have visited. Before the bather is allowed to enter the mineral bath he is obliged to purify himself by a preparatory soap and water ablution, the "bain de proprietie." The water is remarkably clear and transparent, so that every grain of sand at the bottom is distinctly visible, although covered by some three feet of water, through which minute bubbles of gas are continually ascending. The temperature in the principal piscina is ninety-six degrees, and in the other baths it varies from ninety-two degrees to one hundred and three degrees (Fah.)

The arrangements of this establishment, through every portion of which I was conducted by Dr. Haussmann, and the *Badmeister*, are excellent; and the precautions taken to prevent those afflicted with contagious, and horrifying diseases, from bathing in the public baths, are especially deserving of imitation in every similar institution. The prices are very moderate; thus the use of a private *cabinet de bain* costs forty kreutzers, and the public bath thirty kreutzers. Should the bather desire more luxurious accommodation, he may indulge his taste at all prices, up to fifteen francs, at which price he may have the cabinet used by various crowned heads, amongst others last year, by the Emperor of Russia; but the public bath is quite good enough for any person whatever.

We next visited the hospital and baths for the poor, situated in an adjacent building on the river side. This excellent institution, supported by the government, receives patients from every part of Germany, and especially from Würtemburg, most of whom are treated gratuitously; while those who can afford it pay a very small fixed charge towards their maintenance. The baths in this building are exactly on the same plan as those I have already spoken of, and are equally well kept and clean; but the roof not being so very lofty, are considerably warmer, which Dr. Haussmann considers an advantage for this class of patients, most of whom labour under diseases resulting from exposure to cold and wet.

The reputation of Wildbad as a Spa is of very ancient date, and in the reign of Charles the Fifth, by the gratitude of some courtier, who here regained his health, a very curious charter was obtained for this town, which contained a proviso that "all criminals, with the exception of murderers and highway robbers, might here enjoy peace and quiet, undisturbed, for a year and a day."

The geological structure of the country about any watering-place is always of considerable importance, when we wish to study the nature of its mineral springs; and, in the present instance, we find that the surrounding mountains consist of red, ferruginous sandstone and granite, and that the springs issue from clefts in this granite formation.

According to Fehling, the most recent authority, the following is the

ANALYSIS OF WILDBAD WATER.

| | |
|---|---:|
| Sulphate of soda | 0·29 |
| Sulphate of potash | 0·10 |
| Chloride of sodium | 1·80 |
| Carbonate of soda | 0·83 |
| Carbonate of magnesia | 0·07 |
| Carbonate of lime | 0·73 |
| Carbonates of iron and manganese | 0·02 |
| Silex | 0·48 |
| Total | 4·35* |

According to other chemists the total solid ingredients of this water amount only to three and a half grains to the pint, two grains being common salt, the rest chiefly carbonate and sulphate of soda. The gaseous constituents are, however, more important, and abundant, one hundred parts consisting of

> 2·00 carbonic acid,
> 6·54 oxygen,
> 91·56 nitrogen.

These constituents, solid or gaseous, are obviously insufficient to account for the active therapeutic properties of Wildbad water. Some writers ascribe these solely to its temperature, which is ninety-eight degrees, or exactly that of the blood.

With respect to the *modus operandi* of these waters, I had a long argument with Dr. Haussmann, and I here

* *Dr. Seegen's—Handbuch der Heilquellenlehre*, p. 652. Wien, 1862.

subjoin the opinion of that experienced and accomplished physician. His idea is, that the therapeutic action of the Wildbad water is mainly due to the fact that it contains no lime whatever, although it is a very powerful solvent of that base, and therefore, that it acts by dissolving and removing the salts of calcium which exist in excess in the blood and tissues. He also gave it as his opinion, that by increasing the fluidity of the blood it thus facilitates the elimination of morbid materials from the system, and in part also obtains the same result by stimulating the excretory organs, especially the kidneys, and skin. The new springs which have been recently discovered in the site of the clergyman's house, between the Hotels Bellvue and Klumpp, are so nearly identical in composition with those already spoken of, that I need not give any detailed analysis of them. Six springs have been discovered in this spot, and a rude shed has been erected over two of them; the rest having been plugged up until it shall have been decided whether they shall be used for new baths or for drinking purposes. The taste and physical qualities of these sources are in all respects identical with the principal spring, with the exception of their being higher in temperature.

The first effect of the Wildbad baths is a peculiar sense of comfort, or "bien-être," which has been, I think, somewhat exaggerated by most writers. Succeeding to this is a slightly stimulant or exhilarating influence, which, if the bath be too long continued, is followed by a feeling of weariness or lassitude, and a soporific tendency. Therefore the local physicians enjoin their patients to commence with a bath of ten minutes' duration, which may be gradually increased until it at last reaches half-an-hour's immersion, beyond which it will be very seldom, if ever, proper to prolong the bath. After some time the patient will generally experience the symptoms of what I have already had occasion to describe as the Spa fever, or saturation point, and this usually proves critical, and is a precursor of the cure of the ailment for which the invalid has visited Wildbad.

Among the many diseases in which the waters of Wildbad are prescribed with advantage, are old gun-shot wounds, and contractions resulting from this cause; some forms of paralysis, especially of the lower extremities, and when affecting one side only. They should be forbidden, however, in all such cases when plethora is present. These baths are also largely employed in the treatment of neuralgia, sciatica, and some other nervous affections; in hysteria and other diseases peculiar to females, when dependant upon the obstruction of certain functions. They are also useful in many cutaneous affections, among which are included herpes, acne, pityriasis, prurigo, chronic psora, and ringworm of the scalp; in some scrofulous and glandular maladies, and, above all, in chronic gout and rheumatism. In the last-named complaint, it is indeed that Wildbad seems to exercise its most marked curative effect; and more especially is this sanative influence shown in cases of chronic rheumatic arthritis, in which the action of the joint is impaired, or even its form altered, by morbid deposits. Such structural changes are often rectified, and the effused matter absorbed by the combined internal and external use of this, apparently, simple water, when more pretentious remedies have been long tried in vain.

## CHAPTER XIX.

### BADEN-BADEN.

*Its Topography—Its Advantages and Disadvantages—Descriptions of the old and new Towns—Hotels and Lodgings—Number of foreign Visitors—The "Conversation-Haus"—A strange Scene—Society in Baden—The water as "improved"—The Mineral Springs compared with Wiesbaden—Composition of the Ursprung—Medicinal Action of the Water—Observations on its Employment—Its use in cases of Rheumatism and Gout, Nervous Affections, Indigestion, Chronic Catarrh, &c.*

BADEN-BADEN, the *soi disant* " Queen of the Spas," lies twenty-four miles south of Carlsruhe, the capital of the Grand Duchy, thirty-six miles east of Strasburg, and six miles from the Rhine, on the little river Oosbach. The town is built amphitheatre-like, on the descent of the Schlossberg, and in the valley at its foot, which is surrounded by thickly-wooded hills.

The situation of Baden is exquisite; the town is accessible; the hotels and lodging-houses are commodious; the people are civil; the shops are good; the cursaal is the richest, and the mineral waters are the poorest on the Rhine. No watering place shows such evidence of prosperity; and none deserves it less. New streets and houses are rising on every side; the number of visitors is increasing each season; the mineral springs are recommended in almost every class of chronic disease, and are suited for hardly any.

Baden is completely divided into two distinct towns —the old and the new,—the buildings of which are as different as their populations. The former, which is situated on the hill, consists of narrow lanes of quaint, old-fashioned houses, rising in successive terraces, is peopled by the indigenous inhabitants, and contains the mineral sources. The new town occupies the valley I have spoken of, and is formed principally by two streets, namely, the Langen Strasse, or main thoroughfare, which runs parallel with the river, and is the principal business, or shop street; this leads into the Leopold Plaz, whence the Sophien Strasse, a kind of boulevard, runs up to the hospital church. The cursaal, new "Trinkhalle," or pump-room, the public gardens, and numerous villas and boarding houses lie on the south of the rivulet. The modern part of the town is exclusively populated by the foreign element, and it would be difficult to find a dozen large houses here which are not either hotels or lodging houses.

Some idea may be formed of the concourse of strangers to Baden from the number of these hotels and lodging houses, as, in this little town of seven hundred houses, there are no less than forty of the former, and between three and four hundred of the latter.

The season before last, forty-six sovereigns and princes visited this watering place. The following table, abstracted from the *Mercure de Bade* shows the rise of this Spa from a small watering-place to its present position as "Queen of the Spas":—

NUMBER OF VISITORS TO BADEN-BADEN.

| Year. | Number. | Year. | Number. |
|---|---|---|---|
| 1790 | 554 | 1850 | 33,632 |
| 1801 | 1,555 | 1855 | 49,067 |
| 1810 | 2,462 | 1859 | 34,595 |
| 1820 | 5,132 | 1860 | 46,842 |
| 1830 | 10,992 | 1863 | 47,000 |
| 1840 | 20,000 | 1864 | 46,038 |
| 1849 | 14,446 | | |

During the season of 1865 the number of visitors to Baden was upwards of fifty thousand; but owing to the disturbed condition of every part of Germany during the war the number of visitors to this town in the season of 1866 was considerably lessened.

The great rendezvous of all the strangers in Baden is the cursaal or "Conversation-Haus," as it is called, on the *lucus a non lucendo* principle, since conversation is the last thing thought of in these saloons, where the hushed silence which generally reigns paramount, is broken only by the voice of the croupier as he announces the winning colour, or the click of the ball as it spins round the revolving disc. This cursaal always seemed to me by far the worst of the German gambling houses; there is no pretext of any other object—no reading or billiard rooms, as at Homburg or Wiesbáden. In other Spas where gambling exists, it is bad enough, Heaven knows, but still it may be avoided, and invalids may, and do reside for the season without being brought into contact with the gaming table.

But in Baden, gambling is the main characteristic of the place; it is thrust before one prominently at every moment, and it would be impossible to escape its atmosphere in the town; everyone speaks of it, everyone thinks of it, and not a few dream of it. As long as the rooms are open, from eleven in the morning till midnight, they are constantly crowded with players, and from morning till night an equal throng press about the tables in the vain pursuit of easy wealth,— like moths round a flame, although their wings be often singed, yet the fascination is irresistible, and they still rush back to the danger, until, their golden pinions destroyed, they fall into and perish in the fire they have worshipped.

"Poverty," it is said, "makes us acquainted with strange associates;" but assuredly the *auri sacra fames* introduces men into yet more questionable society. Struggling for places round the gaming tables of Baden may be seen noblemen and their grooms, ladies of high rank and spotless reputation and ladies of the "demi-monde," clergymen and blacklegs, all elbowing and pushing aside each other on terms of perfect equality, for a nearer view of the object of their common anxiety—the colour of the card, or the number of the ball, on which sudden accession of riches or great loss, perhaps, depends. The play is on a much higher scale here than at the other Spas, and even the smallest stake allowed is nearly twice as high as at Homburg, while the chance of winning at the roulette is only half, as there are two zeros instead of one. The highest play is at the rouge et noir, at which few stake anything but gold; and I have often seen a player leaving as much as twelve rouléus of Napoleons at the table, and last season I saw one young lad who played habitually, and always put down the maximum stake allowed, on each coupe. Even to those who are resolved not to play, so polluted a moral atmosphere must be very injurious; they become habituated to the sight, and come at last to think that there can be, after all, no such great harm in it. On medical grounds, the state of excitement and tension which gambling produces, is as injurious to the physical health of the player, as on higher grounds it is to the moral.

The Baden Spa is still prescribed by many physicians, and employed by some patients; but out of the fifty thousand visitors whose names figure in the *Kur-liste*, comparatively few use the waters, either internally or externally. Every morning during my residence there, I visited the various springs and the old and new trinkhalles, and seldom met many valetudinarians at the sources. Occasionally, however, I saw a thin muster of invalids at the Ursprung, and always found some thirty or forty water bibbers in the new trinkhalle. Hardly any of the latter, however, had the aspect of sick people, and several of the gentlemen qualified the dose with some drops of a preparation, more potent than Ursprung water, which the very attractive feminine chemist behind the marble counter dispensed from a black bottle.

In composition, this Spa resembles that of Wiesbaden, being a warm saline water, but is much weaker, as there are nearly thirty grains of chloride of sodium, and four cubic inches of carbonic acid gas, more in each pint of the Wiesbaden than in this water. The following is the

ANALYSIS OF THE URSPRUNG, BY BUNSEN.

| | |
|---|---|
| Chloride of Sodium | 16·52 grains. |
| Chloride of Magnesium | 0·09 ,, |
| Chloride of Potassium | 1·25 ,, |
| Phosphate of Lime | 0·02 ,, |
| Sulphate of Lime | 1·55 ,, |
| Sulphate of Potash | 0·01 ,, |
| Carbonate of Lime | 0·88 ,, |
| Carbonate of Magnesia | 0·02 ,, |
| Carbonate of Ammonia | 0·03 ,, |
| Carbonate of Protoxide of Iron | 0·91 ,, |
| Total Solid Contents in a pint of the water. | 21·35 ,, |
| Carbonic Acid Gas, rather less than a cubic inch. | |

The Baden-Baden waters are used both internally and externally, and both methods are generally combined during the course. The cases in which Baden is resorted to are very similar to those in which Wiesbaden is indicated, but which do not require, and could not bear so powerful an excitant as that Spa. During each of my visits to Baden-Baden I met with some patients suffering from a mild form of chronic rheumatism, who seemed to have derived benefit from this remedy. It is also admissible in some forms of neuralgia; and Dr. Edwin Lee records its virtues in instances of "nervous affections of a convulsive nature, such as hysteria, with congestions of internal organs and irregularity in the performance of periodical functions."* It is prescribed with occasional advantage in the treatment of chronic indigestion, marked by heartburn, acidity of stomach, and painful distention after meals; and in irritability of the vesical and intestinal mucous membranes. Dr. Seegen says that Baden is indicated in cases of chronic catarrh of the respiratory organs.† Although loth to dissent from any opinion of this distinguished German authority on mineral waters, I should hesitate to send patients of mine, suffering from chronic bronchial disease, to so variable a climate, especially as I have no faith in any specific action of the water in such cases, unless the disease were clearly connected with the gouty diathesis. It is also employed in scrofulous affections of the glands and skin, and, by the resident physicians, in almost every other disease in the nosology.

When the Baden-Baden waters are prescribed they should certainly be drank at their source on the hill, and not, as is generally the case, in the new Trinkhalle, where the Ursprung water arrives deprived of a considerable amount of the caloric and gas, to which it must owe much of whatever efficacy it possesses.

Before turning away from the German Spas, my

---

\* *The Baths of Rheinish Germany*, p. 168, 3rd edition. London, 1861.
† *Handburch der Heilquellenlehre*, p. 446.

account of which ends here, I may observe that I have deemed it unnecessary in the preceding chapters to make any allusion to the events of the short, but, apparently, decisive war of 1866, between Austria and Prussia, although many of the most important portions of this campaigne were enacted in scenes which I have here had occasion to describe.

Nor have I made any alteration even in my account of those places which, like the Free City of Frankfort or the Duchy of Nassu, for example, have undergone political changes since I visited them before the outbreak of the war; as I have thought it better to retain my account, written on the spot, of these places as they were when I saw them. For the results of that war are yet problematical, and, however it may finally eventuate, and whatever political changes may take place in some of the countries through which I have attempted to guide the pilgrim to the Spas, the Spas themselves will remain unaffected, and little change will be thus produced in the social aspect of the watering places of Germany.

## CHAPTER XX.

#### THE SWISS BADEN.

A painful operation—Situation of Baden—The town—Mineral sources—Their neglected condition—Analysis of the water—Ancient renommè of the Spa—Spa life in Baden in the fifteenth and seventeenth centuries, as described by eye-witnesses—Cases in which this spring is now employed.

Our journey from the German to the Swiss Baden, passing through the ancient historic city of Basle, occupies some pages in my note-book. But warned by the many Spas we have still to visit, and the lessening space there is now left, I have, though, I confess, with much the same feeling with which a surgeon might operate on his own offspring, cut out almost all my remarks on Switzerland. BADEN ON THE LIMMAT is a very ancient small-walled town on the left bank of the Limmat; a considerable portion of the houses being built on a kind of high platform over the river, while the rest lie in the ravine through which the stream flows. At the south end of the place a very curious, antique covered bridge crosses the river, and at the opposite extremity are situated the mineral baths and springs. The principal source issues in a small irregular place formed by the hotels, and immediately in front of the Switzer Hoff hotel. Nothing can be more uninviting than the appearance of this fountain, which seems to be the receptacle for all the rubbish from the adjacent houses, and the smell of the decomposing vegetable matter rises distinctly over the strong sulphurous odour of the water. The spring itself which is warm and very gaseous, has a disagreeable saline and sulphureted taste.

ANALYSIS OF BADEN MINERAL WATER—By Löwig.

| | | |
|---|---|---|
| Sulphate of soda | 2·218 | grains. |
| ,, magnesia | 2·442 | ,, |
| ,, lime | 10·860 | ,, |
| Chloride of potassium | 0·711 | ,, |
| ,, sodium | 13·042 | ,, |
| ,, magnesium | 0·566 | ,, |
| Carbonate of lime | 2·599 | ,, |
| ,, magnesia | 0·152 | ,, |
| ,, strontium | 0·005 | ,, |
| Fluoride of calcium | 0·016 | ,, |
| Chloride of calcium | 0·719 | ,, |
| All other ingredients | 0·013 | ,, |
| Total solid contents | 33·343 | ,, |

Baden was frequented as a watering place by the Romans, and numerous relics of this bath-loving people have been found in the environs of the springs. In the middle ages Baden was no less resorted to than at the present day, and Poggio Bracciolini, the celebrated Roman courtier, scholar, and antiquary of the fifteenth century, has left, in his correspondence with Nicolo Niccoli, a most interesting sketch of a fashionable watering place of that day, in his " Account of the

Swiss Baden," when he visited that spa in 1415. "I write to you," he says, "from these baths, to which I have now come, to try whether they can remove an eruption which has taken place between my fingers; to describe to you the situation of the place, and the manners of its inhabitants, together with the customs of the company who resort hither for the benefit of the waters. Much is said by the ancients of the pleasant baths of Puteoli, which were frequented by almost all the people of Rome. But, in my opinion, these boasted baths must, in the article of pleasure, yield the palm to the baths of Baden. . . . Baden is a place of considerable opulence, situated in a valley surrounded by mountains, upon a broad and rapid river, which forms a junction with the Rhine, about six miles from the town. About half-a-mile from Baden, and on the bank of the river, there is a very beautiful range of buildings, constructed for the accommodation of the bathers. Those buildings form a square, composed of lodging houses, in which a great number of guests are commodiously entertained. Each lodging-house has its private bath appropriated to its tenants. The baths are, altogether, thirty in number. Of these two only are public baths, which are exposed to view on every side, and are frequented by the lower order of people, of all ages, and of each sex. I admired the simplicity of these people, who take no notice of these violations of propriety, and are totally unconscious of any indecorum. The baths belonging to the private lodging-houses are very commodious. They, too, are resorted to by males and females, who are separated by a partition. In this partition, however, there are windows, through which they can converse with each other. Above the baths are a kind of galleries on which the people stand who wish to see and converse with the bathers; for every one has free access to the baths, to see the company, to talk and joke with them. The bathers frequently give public dinners in the baths, on a table which floats on the water. Our party received several invitations. I paid my share of the reckoning; but though I was frequently requested to favour them with my company I never accepted the summons—not through modesty, which would, on these occasions, be mistaken for rudeness, and want of good breeding, but on account of my ignorance of the language, for it seemed to me an act of folly in an Italian, who could not take any part in conversation, to spend all the day in the water, employed in nothing but eating and drinking. . . . They go into the water three or four times every day, and they spend the greater part of their time in the baths, where they amuse themselves with singing, drinking, and dancing. In the shallower parts of the water they also play upon the harp. . . Besides money, garlands, and crowns of flowers are thrown down, with which the ladies ornament their heads while they remain in the water. As I only bathed twice a-day, I spent my leisure time in witnessing this curious spectacle, visiting the other baths, and causing the bathers to scramble for money and nosegays; for there was no opportunity of reading or studying. The whole place resounded with songs and musical instruments, so that the mere wish to be wise were the height of folly; in me especially, who am not like Menedemus in the play, a morose rejecter of pleasure, but one of those who take a lively interest in everything which concerns their fellow mortals. . . . There is a large meadow behind the village, near the river. This meadow, which is shaded by abundance of trees, is our usual place of resort after supper. There the people engage in various sports—some dance, others sing, and others play at ball, but in a manner very different from the fashion of our country. For the men and women throw in different directions a ball filled with little bells; when the ball is thrown they all run to catch it, and whoever lays hold of it is the conqueror, and again throws it at somebody, for whom he wishes to testify a particular regard. When the thrower is ready to toss the ball all the rest stand with outstretched hands, and the former frequently keeps them in a state of suspense by pretending to aim, sometimes at one, and sometimes at another. Many other games are here practised, which it would be tedious to enumerate."\*

Fynes Moryson, an old English traveller, gives the following description of the baths of the Swiss Baden,

---

\* *The Life of Poggio Bracciolini. By the Rev. William Carpenter.* pp. 69—76. Liverpool, 1802.

as they existed in 1617. These baths, he says, "are famous for medicine, and are in number thirty, seated on each side the brooke, which divideth them into *Bethora*, the great and the little. In the great, divers baths are contained under one roof of a house, and without the gate are two, common to the poore. These waters are so strong of brimstone as the very smoak warmeth them that come neere, and the waters burn those that touch them. Of these, one is called the Marques Bath, and is so hot as it will scald off the haire of a hogge. The waters are so cleere as a penny may be seen in the bottome, and because melancholy must be avoided, they recreate themselves with many sports, while they sit in the water; namely, at cards, and with casting up and catching little stones, to which purpose they have a little table swimming upon the water, upon which sometimes they doe likewise eate. These baths are very good for a cold braine, and a stomach charged with rhume; but are hurtful to hot and dry complexions, and in that respect they are held better for women than men."*

At the present time Baden-on-the-Limmat is frequented by few except Swiss valetudinarians, or invalids from the neighbouring German states, during the season, which lasts from June to September. It is a quiet, and somewhat "*triste*," though beautifully-situated

---

* *An Itinerary written by Fynes Moryson, Gent, Containing his Ten Years' Travels, &c.* Part 1st, Book 1st, p. 26, folio. London. 1617.

watering place. These baths and waters are principally employed in the treatment of some obstinate skin diseases, in secondary and tertiary syphilis, and also in chronic rheumatism and gout, with exudation into the joints.

## CHAPTER XXI.

### SCHINZNACH, WILDEGG, AND PFEFFERS.

Route from London to SCHINZNACH—History and Description of this Spa—Composition, Properties, and Medical Uses of the Sulphurous Water—The Iodated Spring of WILDEGG—Medico-chemical Observations on it—Journey from Zurich—The Lakes of Zurich and Wallenstadt—Ragatz and its Thermal Establishment—Walk from Ragatz to PFEFFERS—Description of this Watering Place—Its situation and romantic Scenery—Remarkable position of the Hot Springs—History of Pfeffers—Analysis of the Springs—Their Character and Uses—Accommodation and Inducements for Invalid Visitors.

RETRACING our steps towards Basle, half-an-hour's journey by train brought us to SCHINZNACH, which is fast becoming the most frequented Spa in Switzerland. As at many other Swiss watering places, at Schinznach there is neither town nor village, but merely a vast bathing establishment, in which those who go through a course of the water must live in community together, completely isolated from the rest of the world.

This Spa may now be easily reached from London, either by Paris and Strasbourg, or by the Great Luxembourg Railroad, *via* Basle, in two days. The situation of Schinznach is eminently calculated, by the quiet beauty of the scenery, to serve a hypochondriac, and by amusing and turning his thoughts from their accustomed channel on the beauties which nature here presents, to divert his attention from his imaginary complaints.

The bath house, which stands on the right bank of the Aar, consists of two parts, which form a crescent, united by a range of buildings at its extremities. The older part of this building dates from the year 1695, and the newer portion was erected in 1842, and both together contain accommodation for about four hundred visitors, and no metropolitan hotel can exceed the excellency of the internal arrangements and conveniences of this establishment.

Behind the main building, and commanding an exquisite view of the valley of the Aar, is the *Hôpital*, where accommodation is provided for seventy-six poor patients, who pay two francs and a half per diem, for lodging, board, and treatment, and the proprietor assured me that by this class of patients he had lost upwards of ten thousand francs. The care taken of them I was glad to observe, however, seems excellent, and the situation of their residence is decidedly superior to that of their more opulent invalid brethren.

The principal sulphurous spring is immediately behind the old building, and rises through a wide well, some twenty-eight feet in depth. The odour is strongly sulphurous, and the flavour is a compound between the washings of a gun barrel and weak brine. The temperature, which I ascertained with considerable difficulty, by lowering myself into the well, is ninety-two degrees.

Schinznach is the most strongly sulphurous Spa in Switzerland, and is moreover stronger than any mineral water of the same class in Savoy or Rhenish Germany.

The following is the most recent analysis of this source :—

| | |
|---|---|
| Chloride of Sodium | 5·001 Grains |
| Chloride of Potasium } Chloride of Ammonia } | 0·063 ,, |
| Sulphate of Soda | 0·919 ,, |
| Sulphate of Lime | 4·886 ,, |
| Sulphate of Magnesia | 2·052 ,, |
| Carbonate of Lime | 1·086 ,, |
| Carbonate of Magnesia | 0·063 ,, |
| Alumina | 0·045 ,, |
| Silicic Acid | 0·086 ,, |
| Total Solid Ingredients | 14·201 ,, |
| Sulphureted Hydrogen Gas | 1·268 Cubic in |
| Carbonic Acid Gas | 1·886 ,, |
| Nitrogen | Traces. |

The most important constituent in this water, in a therapeutic point, is the sulphureted hydrogen gas, the physiological effects of which vary according to the dose,—a small dose being a tonic, while a large one is a powerful stimulant. It is employed both internally and externally, but its principal use is in the latter mode. When used internally, it is necessary to allow it to remain exposed to the air for some time after being drawn from the source, for the purpose of admitting of the escape of a large proportion of the gas it contains. For it would be highly dangerous to use a fluid so strongly charged with sulphureted hydrogen and carbonic acid gas as this is when first taken from the spring.

When drank with the precaution I have mentioned the Schinznach water acts as a stimulant and resolvent; it excites the activity of the gastro-intestinal mucous membrane, increases the secretions from this canal, accelerates the pulse, and determines to the skin. Dr. Amsler [*] and other local authorities contend that this mineral water has a peculiarly stimulant effect upon the bronchial and pulmonary mucous membrane, and that it induces, in a special manner, the secretion of the lungs and bronchial air passages, and that, therefore, it is contra-indicated, and likely to be hurtful whenever there is any tendency to inflammatory action, or even irritation in those parts.

The principal benefit, however, to be derived from Schinznach water, is that produced by its strongly stimulating action on the skin, the capiliaries of which become enlarged, and even to a certain extent actively congested under a course of the baths and waters which generally produce a specific cutaneous eruption, regarded as a proof that the patient's system is under the influence of the remedy. This point is insisted on as of great importance, and is watched for with the same care that we examine the gums of a patient under a mercurial course.

The cutaneous diseases over which the baths seem

[*] Les Bains de Schinznach. Par le Amsler, p. 63. Lenzbourg, 1854.

to exercise most influence are eczema, ringworm, sycosis mentis, scabies, psoriasis, chronic urticaria, and pityriasis. In comparatively few cases of these complaints, however, is it necessary to send our patients to this remote Swiss valley in search of health, as the pharmaceutical means at our hand are sufficient to treat successfully such ailments. But whenever these fail, as they will sometimes do, we have here a remedy which seems to exercise its curative action most potently, in those cases which physic alone cannot cure.

Scrofula, next to skin diseases, is the malady which brings most patients to Schinznach. The stimulant action of the water on the glandular system renders it peculiarly adapted for scrofulous glandular diseases, either external or mesenteric. In scrofulous ophthalmia, its action is well marked, though in such cases it cures the ophthalmia indirectly by its action on the system, correcting the faulty constitutional state, of which the local complaint is but a symptom. Rachitis, scrofulous diseases of the joints, caries, and nervous and other affections resulting from the scrofulous diathesis, are said to be equally curable by the Schinznach Spa.

In chronic rheumatism, I was assured that this water acted almost as a specific, and cases were detailed to me, in which patients who arrived here so crippled by that disease as to be almost incapable of moving themselves, walked down after a few weeks to the train, and went back cured, or greatly relieved, by a course of these baths and waters. In diseases of the digestive organs, marked by slow digestion and habitual constipation, and in diseases occasioned by deficient or irregular menstruation, and in anœmia, the stimulant properties of the sulphurous spring are occasionally beneficial. These very properties, however, it should be observed, render Schinznach water an unsuitable and even dangerous remedy for any patient, be his disease what it may, who is of an excitable, sanguine temperament, and full, plethoric habit of body. And even when the plethora or congestion is local, in the lungs, or uterus for instance, and in every case in which any tendency to cerebral fulness, or apoplexy is feared, and in hypertrophy of the heart, or in aneurismal cases, the incautious use of this Spa, if only once tried, might lead to a fatal result.

The season lasts from May to September, inclusively, and the mode of using the water is the same as at other strong sulphurous baths, externally in douche, vapour, and other baths; and internally in doses of from one to three small glasses, morning and evening.

Before concluding my remarks on Schinznach, I have to express my obligations to the proprietor, for the courtesy with which he afforded me the fullest information and assistance in investigating this Spa.

Close to Schinznach is WILDEGG, where a very

strongly saline iodated spring was discovered in 1830, the water of which is now largely exported.

The source of Wildegg rises through an artesian well, some three hundred feet deep, which furnishes so small a supply, that it can hardly fill fifty small bottles daily. Wildegg is peculiarly interesting as belonging to a class of mineral waters of which we have very few examples, namely, the Iodated and Bromated Spas. According to Dr. Laué, of Wildegg,* the following is the analysis of this spring:—

| | |
|---|---|
| Iodide of Sodium | 0·218112 grains. |
| Bromide of Sodium | 0·236544 ,, |
| Chloride of Sodium | 80·236800 ,, |
| Chloride of Potassium | 0·039936 ,, |
| Chloride of Lime | 1·980672 ,, |
| Chloride of Magnesium | 12·451584 ,, |
| Chloride of Strontium | 0·152832 ,, |
| Hydro-chloride of Amonium | 0·049152 ,, |
| Sulphate of Lime | 14·172672 ,, |
| Nitrate of Soda | 0·339456 ,, |
| Carbonate of Lime | 0·583680 ,, |
| Carbonate of Iron | 0·061400 ,, |
| Silicic Acid | 0·030720 ,, |
| Total solid contents of 16 ozs. | 110.553600 grains. |
| Carbonic acid gas | 2·36 cubic inches. |

Small as the amount of iodine and bromine in the Wildegg water may seem, when compared to the great strength of its other saline constituents, it is to these salts that the Spa owes its reputation as a water of remarkable efficacy in the treatment of chronic glandular and scrofulous diseases; the iodine

* Etudes sur les Eaux Minerales de Schinznach et de Wildegg. Par A. Hermman, p. 22. Zurich. 1858.

and bromine stimulating the absorbent system, and thus dissipating and removing chronic indolent tumours. The water must be administered very cautiously, however, for if used too long or in too large doses, it occasions all the symptoms of iodism. It is generally combined with a course of the Schinznach water and baths, and is given in doses of from two to four ounces night and morning.

Comparatively few tourists in Switzerland visit PFEFFERS, though many pass through the adjoining village of Ragatz, and yet this is perhaps the most interesting spot in Helvetia, as well as one of the most remarkable Spas in Europe. On our journey to this watering place we left Zurich early in the afternoon and arrived at Ragatz late at night. The journey from Zurich affords illustrations of every variety of Swiss scenery. Immediately after leaving that city, the railway passes through a highly-cultivated and densely-populated country, in which the small farms are so skilfully tilled, and the peasant cottages so tastefully built and cleanly, that we might imagine ourselves to be in the scene the poet's fancy pictured in those " merrie days "—

"Ere England's griefs began,
When every rood of ground maintained its man."

At Rapperschwyl, the railroad approaches the Lake of Zurich, through a part of which it passes on a narrow causeway for some distance. This portion of the lake is surrounded by vineyards and wooded hills, sloping down to the water's edge, upon which are to be seen numbers of small sailing vessels and steamers, and presents a scene as animated and bustling as those Japanese waters, on which the business of whole cities is carried on, being perfectly different in character from any other Swiss lake.

Following the valley of the Linth, the road turns away from the smiling inland sea of Zurich, and approaches the wild and romantic lake of Wallenstadt, along the whole length of which it runs, either on the verge of the water, or in tunnels through its rocky borders. None of the Swiss lakes seem to me comparable in scenery to the wild and almost savage sublimity of this sheet of water, dark as—

> "That lake, whose gloomy shore,
> Skylark never warbled o'er,"

surrounded on the northern side by lofty granite walls rising straight from the water, and not affording the narrowest ledge from the water below to the precipices above, upwards of three thousand feet from the lake.

After a short stay at Wallenstadt, the small town at the extremity of the lake, under the "Seven Electors," as the peaks behind it are called, we resumed our journey, and passing by the embankment which divides the "Mighty Rhine" from Lake Wallenstadt, and which promises at no distant period to give way and turn the Rhine into the Lake of Zurich, we arrived at Ragatz.

This village, which is now one of the most frequented watering places in Switzerland, consists of a long irregular street, leading from the railway station, and crossed at its extremity by a shorter *strasse*, in which is the hotel, and bath house. The population does not probably exceed one thousand inhabitants, and excepting the baths, and the vista it commands of the neighbouring scenery, the village contains little to attract tourists. Its position, at the mouth of the wild gorge from which the Tamina issues, and the view from the hotel of the wide amphitheatre-like valley, surrounded with snow-covered mountains before it, are, however, certainly highly picturesque.

The principal hotel, the Ragatz-Hoff, is a large and handsome building, with a commodious bath establishment attached to it. But as it is supplied from the source of Pfeffers, some miles distant, and as the curative effects of this water are, probably, more connected with its thermal condition than with its chemical composition, the former must be considerably impaired by its long transit through the wooden pipes. There can therefore be no possible advantage in using the baths at Ragatz instead of at their source.

From Ragatz, a steep, winding road, immediately behind the hotel, brought us to the cliffs which overhang the Tamina, along the left bank of which the path is cut through the rocks.

Gazing down the dark rocky ravine, we saw the Tamina foaming along with tremendous velocity, tumbling over huge boulders, and forming miniature cataracts on its way to the distant Rhine. The road here and there descends to the level of the water and then rises to the edge of the precipice above it. In some places it is carried through short tunnels in the rock; and nowhere is the route devoid of picturesque beauty, and often passes through wild and sublime scenery. Near to Pfeffers this road was recently the scene of a fatal accident, by which three ladies, two of whom were Irish, and the other a German, lost their lives. These ill-fated travellers, who were journeying on a pleasure tour, were returning from Pfeffers to Ragatz on the 3rd of July, 1866, when the horse suddenly took fright and precipitated himself, with the carriage and passengers, from the narrow roadside, which is unprotected by any barrier, into the Tamina, a distance of about thirty feet, where they perished. The driver, who was thrown from the box on to the road, being the only one of the party that escaped to tell the sad tale.

A couple of hours' brisk walking brought us to our destination at Pfeffers. The old convent or bath house, is built on a narrow ledge of rock over the torrent, and is overshadowed by the opposite precipice, which, rising four or five hundred feet high, keeps the house in perpetual shade. The building is a long, narrow edifice, six stories high, and at the time of our visit, when the season was nearly over, looked exactly like a deserted cotton factory.

The bath house of Pfeffers is the property of the government of the canton, and contains accommodation for about three hundred visitors, and though this plain, unornamented building affords little luxury to its inmates, yet their comforts seem well cared for, and the very reasonable and fixed prices charged for board, lodging, and the use of the waters, as well as the civility of the manager and attendants, render it preferable in these respects to the rival establishment at Ragatz. The prices at Pfeffers for rooms vary from one to two francs a day—breakfast one franc, dinner from two to three francs, and baths one franc. The rooms on the ground floor are given up to the use of poor patients, who are received and treated gratuitously.

By a narrow wooden bridge we crossed the Tamina, on the western end of the convent, and got on a narrow pathway cut out of the rock. The ravine along which we had walked to Pfeffers, here suddenly contracts to a crevasse not thirty feet across, and the precipices on either side approaching and sloping towards each other, form a limestone roof some four hundred feet high, and thus enclose a long cavern, lighted by the few rays that find their way through the fissures where the rocks above meet. Along this cavern the pathway to the thermal sources passes for nearly half-a-mile, midway between the roof and the abyss through which the torrent below rushes; the passage being supported partly on a narrow ledge of rock projecting over the Tamina, and in part on stakes driven into the marble walls of the crevasse. At the extremity of this pathway are situated the thermal sources to which Pfeffers owes its fame. They both rise within a few yards of each other, at the bottom of a cavern on the right bank of the Tamina. The new source which was discovered in October 1860, was the first we examined. It is approached by a long tunnel cut through the solid rock, and so filled with steam that the torches we carried gave scarcely a glimmer of light, but guided by the voice of our conductor, and almost parboiled, we reached the hot spring, which has neither taste nor odour, is perfectly limpid, has the specific gravity of common water, and issues from the fountain at the temperature of 99 deg.

The old source adjoins, and was first employed, as the inscription over it informs us, in 1038. It is contained in a similar grotto to the last-named one, but is reached by a much shorter passage. The temperature of the water was, I found, one degree lower than the new spring, but in all other respects it seems identical with it.

On the other side of the torrent, which is here crossed by a very narrow wooden bridge, is a remarkable excavation, or grotto, in the form of an amphitheatre, some thirty feet wide, by fifty in depth. This our cicerone informed us was the grotto of St. Madeleine, once a celebrated shrine for pilgrimages.

Returning by the same suspended road, and holding

on by the wooden pipes which convey the water from its sources to the bath houses of Pfeffers and Ragatz, we came back to the convent.

The thermal source of Pfeffers was first discovered in 1038, by a huntsman in quest of quarry, who, alarmed by seeing steam rising from the ground before him, turned back from the chase, and communicated his discovery to the monks of the adjacent convent. By them it was utilised for the benefit of the poor of the district, and its fame gradually spreading, the first thermal establishment was opened in 1242.

The original bath-house consisted of a small wooden building, suspended on beams of timber over the abyss of the Tamina, and communicated with the source immediately below with a ladder, and in this aerial abode the patients remained confined during the course.

In 1630, this house, or rather cage, was destroyed by fire, and the new monastery at the mouth of the cavern being just completed, patients were received there—the poor gratuitously, and the rich for whatever offerings they might choose to make on their departure. Thus the baths of Pfeffers flourished till 1838, when the government of the canton dispossessed the monks, and turned the monastery into an hydropathic establishment.

From that time Pfeffers became, year by year, less frequented by invalids, who now seem to prefer the modern rival establishment at Ragatz. The waters of Pfeffers belong to the same class of mineral springs as

Gastein and Wildbad, which possess hardly any chemical ingredients, and depend for their action on their temperature. The composition of Pfeffers water, according to the most recent authority, is as follows:—

ANALYSIS OF PFEFFERS SPA ACCORDING TO PAGENSTECHER.

| | |
|---|---|
| Sulphate of Soda | 0·242 |
| Sulphate of Potash | 0·004 |
| Chloride of Sodium | 0·208 |
| Sulphate of Lime | 0·027 |
| Carbonate of Lime | 0·910 |
| Carbonate of Magnesia | 0·147 |
| Other Salts | 0·148 |
| Total solid contents | 1·792 |

The feeble mineralization of the Pfeffers waters explains the impunity with which the large doses commonly used can be taken. But this does not, however, explain how it is that while tepid water, even in small draughts, causes sickness or nausea, patients with weak stomachs can drink from ten to twelve small glasses of this tepid Pfeffers water a day, with positive increase of appetite.

The principal use of the thermal sources of Pfeffers is in the baths, in which the patients sometimes remain for a considerable time, though now they no longer remain as formerly, when, as an old author assures us, they stayed for whole days together in the baths: "Multa dies noctesque thermis non egrediuntur; sed cibum simul et somnium in his capiunt."

The average duration of each bath at Pfeffers at

present is reduced to about twenty minutes. The same peculiar sense of *bien-etre* is ascribed to them as to the Wildbad baths, and it is certain that they do exercise a remarkably sedative, but not depressing influence on the system. After a few days, a slight febrile reaction comes on, during the course of the baths, and this is regarded as critical, and usually, but by no means always, foretells a cure.

Long, indeed, would be the list, if I merely enumerated the diseases which the special writers on Pfeffers say may be cured by its waters. I shall, however, now only allude to those ailments which my own experience leads me to think are most susceptible to the therapeutic influence of this Spa.

Foremost amongst these complaints, is dyspepsia and gastralgia, and it is remarkable in what large doses even the most irritable stomach tolerates the water, and how rapidly the best effects—diminution of pain, regular alvine action, and increased appetite—often follow its use.

Nervous and spasmodic affections are frequently benefited by Pfeffers Spa, which soothes and tranquilizes the nervous system in a special manner, and cases of intractable neuralgia, sciatica, nervous headache, and similar complaints, are sometimes cured by a few weeks' use of these baths, and the internal administration of the waters.

The same remarks may be made of hysteria, chorea, and some other obscure nervous maladies. Also in renal and vesical complaints, such, for instance, as catarrh of the bladder; even when the secretion is attended with pain, the passage through the system of so large a quantity of bland fluid as is daily drank at Pfeffers is likely to be attended with the best effects.

## CHAPTER XXII.

AIX-LES-BAINS AND VICHY.

Situation and Description of Aix—Thermal Establishment—Remains of Roman Baths—Casino—Hospital—Thermal Caves—History of the Spa—Number of Visitors—The Mineral Waters—Diseases in which they are prescribed—The Springs of Marlioz—Journey from Aix by the Rhone—Vichy—Account of the Town and its Environs—Accommodation for Invalids—The Bath-houses, Ancient and Modern—Geological Formation of the Country—General Observations on the Mineral Waters—Analysis and particular Account of each Spring—Therapeutic Action of Vichy Water—Its Medicinal effects and the Diseases in which it is employed—Cusset, and its Alkaline Spa.

Aix-les-Bains, or Aix-en-Savoie, as it is still called, to distinguish it from Aix-en-Provence, is beautifully situated, ten miles from Chambery, on the Geneva road, and about three quarters of a mile from the lake of Bourget.

Since the annexation of Savoy with France, the facilities for reaching Aix from Paris have been greatly increased, and it is now within fifteen hours of Paris by rail, *viâ* Mâcon.

The town lies half a mile from the railway station, and is approached by a handsome avenue of chestnut trees. It contains a resident population of between two and three thousand inhabitants, which is annually increased by some eight thousand visitors during the season, and nearly every house is then converted into a lodging-house or hotel. While the season lasts, and the weather continues fine, Aix, although not containing much in itself to attract the visitor's admiration, except the springs, and the ancient and modern thermal establishments, has as animated and lively an appearance as any German watering-place. The town, baths, and Casino are then crowded with invalid visitors, and the environs are alive with their more robust companions, boating on the lake, or mounted on donkies, which are here in universal requisition by all classes and ages alike, climbing the surrounding hills. But when the season is over, nothing can present a more cheerless aspect than the deserted streets and abandoned houses of Aix.

The main feature of Aix-les-Bains is its thermal establishment, which is one of the most complete in Europe. This is a handsome granite building on the hill, a little above the town, and is supplied by the two mineral springs which have their sources within the mountains above it. The bath house consists of two stories, and was originally constructed in four divisions, and at different times; these were joined together and modernized a few years since.

The total number of baths contained in the establishment is three hundred, and includes every variety, from the simple reclining bath to the most complicated local douche, or "pulverized water" bath. The swimming baths and the *vaporarium*, on the model of that at Ischia, are especially deserving of examination. Nearly opposite, and on the same plain as the grand bath house, behind the *Pension Chabat*, are the remains of the extensive ancient Roman baths. The entrance to these is through a coal cellar, under the house, from which a steep narrow passage brought us to a large vaulted chamber, supported on brick columns. This seems to have been used as a swimming bath, and was supplied by a square canal, now stopped up, leading into the mountain.

Besides this, the remains of vapour, reclining, and other baths may be traced. Indeed, the houses in this part of the town are for the most part built from the Roman remains which abound on every side, and some of which—for instance, the handsome triumphal arch of Campanus—still exist in perfect preservation. Some traces of the temple of Diana may also be seen, near the Casino, a large and handsome building in the valley, immediately below the main street. This Casino was originally fitted up and used for gambling, and is decorated in the same style as the German Cursaals. But since the annexation with France, where gambling is prohibited, it has been converted into a kind of club and concert hall, to which visitors find an easy admission.

Opposite to the thermal establishment is the hospital, originally founded in 1813 by Queen Hortense, subsequently endowed by the charity of an English gentleman residing at Lausanne, and lately rebuilt by the French government. A little below the hospital is the church, a very ancient, but unimposing edifice, and here I heard one of the most eloquent discourses from the village curé that it has been my good fortune ever to listen to.

Foremost among the lions of Aix, are the recently-discovered caves in the mountain behind the town. These up to a few years ago, were vast receptacles filled with hot mineral water, and were drained at vast expense by the government, and thus the supply of thermal water was not only greatly increased, but rendered regular and equal. They are entered by a small aperture in the mountain, leading into a long tunnel cut through the solid rock. Having proceeded along this for some distance, we turned up a steep crevice through the limestone, and, ascending on hands and knees, struggled through a narrow entrance into a lofty circular chamber, with a vaulted dome, which the torches rendered visible. Thence we came into a series of other caverns, some similar to the last, others like vast Gothic churches, with pointed roofs, supported on limestone columns, and many others, in which the limestone, eaten away by the action of the hot sulphurous water for successive centuries, had assumed all kinds of fantastic shapes and resemblances. The exploration itself into the remoter caverns would be no easy task for an invalid or a lady. In some places we had to crawl at full length under projecting rocks, which threatened to fall on our heads, in others we passed in the same uneasy fashion through pools of water, and even in the parts which are generally visited by tourists, the damp, close air, and the water constantly dropping from the stalactites overhead, are calculated to injure valetudinarians, and should prevent this class of travellers from visiting these caverns. Nor were my companions and myself sorry when we returned to the tunnel we had considered so dark and dismal before we had seen the caves. At the end of this passage is the *Source de St. Paul*, or *d'Alun*, as it is called. This arises from a large square well, some twelve or fifteen feet deep, and is covered with a wooden structure, on opening which a dense volume of steam rushed out, filling the tunnel. When this had cleared away to some extent, I descended, and proceeded to examine the water. The result of this examination will be found in another page.

From the earliest time that we have record of the medicinal use of mineral waters, Aix-les-Bains seems to have been one of the most celebrated Spas known to the ancients, and especially to the Romans, who named these springs Aquæ Gratianæ. They sank, however, into comparative disrepute, and again grew into favour within the last forty years. In 1842, there were about two thousand visitors, in 1861 there were nearly six thousand, in 1864 there were upwards of eight thousand, and in 1865 and 1866 the number of foreign invalids residing in Aix-les-Bains continued to increase with the same proportionate rapidity.

The springs of Aix-les-Bains belong to the class of warm sulphurous waters. There are two sources— one the sulphurous, the other miscalled the *source d'Alun*, which, however, contain hardly any trace of alum, and is properly designated the *source de St. Paul*.

ANALYSIS OF AIX-LES-BAINS SOURCES ACCORDING TO DR. SEEGEN.

| A Litre of Water contains (in French Grammes) | Sulphurous Source. | Alum Source. |
|---|---|---|
| | Gramme. | Gramme. |
| Carbonate of lime | 1·1803 | 1·2384 |
| Carbonate of iron | 0·0387 | 0·0774 |
| Chloride of Calcium | — | 0·4644 |
| Chloride of Magnesium | 0·1548 | 0·1548 |
| Sulphate of Lime | 0·4257 | 0·6966 |
| Sulphate of Magnesia | 0·7353 | 0·2322 |
| Sulphate of Soda | 0·3483 | 0·2322 |
| Baregine or Glairine | An unascertained quantity. | An unascertained quantity. |
| Total | 2·8831 | 3·0960 |
| Temperature | 115 deg. (Fah.) | 117 deg. |

The alum water is that generally taken internally, being less nauseous and easier of digestion than the sulphurous source. But comparatively little internal use is made of the waters at Aix. It is in the baths, and especially in the douche baths, that their efficacy is most often proved.

The diseases in which the baths of Aix are commonly recommended are—chronic rheumatism, especially when enlarging and disabling the joints, rheumatic gout, some forms of local paralysis, when there is no tendency whatever to cerebral congestion, chronic ulcers, and old and painful wounds that either break out afresh, or become painful at recurrent periods. Baron Despine states, that the inhalation rooms at Aix and Marloiz are frequently used with great benefit in chronic catarrh, and humoral asthma.* But after all, the thermal sulphurous baths of Aix seem best adapted for the treatment of chronic and intractable skin diseases, especially syphilitic and mercurial eruptions, or when it is suspected that those poisons lurk in the system, and give rise to obscure and anomalous symptoms. The local douches are also applied with great benefit to diseases of the eye and ear, and the so-called "pulverized water," is used with advantage in cases of clergyman's sore throat and in ozena.

Three quarters of a mile from Aix-les-Bains, on the road to Chambery, is the new Spa of MARLIOZ, where a *sale de inhalation*, and pump room, has been constructed within the last ten years. This building is a long, narrow pavilion, consisting of an open portico, in which the two mineral sources flow from an artificial rocky grotto; and a couple of rooms—one for gentlemen, the other for ladies, where the water is forced up and reduced into a fine spray, which is inhaled for various periods by the patients.

* "The Baths of Aix-en-Savoy." By the Baron Despine, M.D., p. 12. Paris. 1861.

Marlioz is a cold, strongly sulphurous spring, containing traces of iodine. Its action is stimulant, and it is employed internally, and also in local douches, as well as by inhalation. By the physicians of Aix it seems regarded as almost a specific in pulmonary diseases, especially in chronic bronchitis; in the early stage of consumption occurring in persons of a leucophlegmatic temperament, and even when hœmoptysis has appeared. And also, though it seems hard to conceive how one remedy can cure so many complaints, often very unlike each other in all but their situation in the chest, it is said to be equally useful in chronic catarrh, asthma, and angina.

Our route from Aix to Vichy by steamer, through the Lake of Bourget, and down the Rhone to Lyons, and thence to Vichy, led us across a most interesting part of France, and did my space permit, gladly would I expatiate on the pleasant days this journey, by steam, by coach, and on foot, afforded me. For, sauntering as we did slowly from place to place, during an autumn ramble, we in turn tried each of these modes of peregrination. But I must pass by these reminiscences, and hurry on to our destination, where we arrived before the last invalid had left for the season.

VICHY, which is the Spa par excellence of France, is situated on the Allier, in the department of the same name, ninety eight miles from Lyons, and within eight hours' drive of Paris, by the Lyons Bourbonnais railway.

Vichy has a resident population of about two thousand inhabitants, and is divided into two distinct parts, viz., *Vichy les Bains* and *Vichy la Ville*, which are separated by the park. The former is the new town about the railway station, and consists of two or three long, handsome streets of hotels and lodging-houses, and in this part of Vichy almost all the visitors reside.

A very handsome church has been just built in this quarter, and was opened last season. This part of the town is thronged during the season, and contains a superb Casino, the hotels, numerous shops, such as may be found in all watering-places, with a few billiard-rooms and cafés, and when the season is over is almost deserted, and all the hotels and lodging-houses are closed till the following year.

*Vichy-la-Ville*, the old part of the town, is, if possible, more crowded after than during the season, being filled by a poorer class of patients, who, coming hither in the hope of being cured of some disease, and not for fashion, wisely consider that they may expect more attention, and equal benefit, after the beau monde have returned to Paris. Vichy-la-Ville bears unmistakeable traces of its antiquity, in the style of the houses, and the dark, narrow, tortuous streets, or rather lanes, of which it is composed.

Neither the new nor the older part of Vichy present much deserving of special notice, excepting the mineral springs, the new Casino, and the thermal establishment. One of the principal advantages of Vichy, over some other watering places, is the fact that the invalid is here enabled to obtain from the local physicians the highest class of medical assistance. Among the numerous physicians who practise in Vichy during the season, I may make special mention of Dr. Barthez, author of a well known work on this Spa,* as one of the most eminent. The accommodation for visitors afforded by between seventy and eighty hotels, and almost as many boarding establishments as there are houses in the town, is generally excellent, but extremely expensive, especially during the month which the French Emperor and court annually pass there. Indeed Vichy owes much of its present prosperity as a fashionable watering place to the regular annual visits of the Emperor. Thus, before the Emperor patronized these waters, in 1854 for instance, the total number of invalids that resided here during the season was between six and seven thousand, whilst last year the number of valetudinarians who followed the Imperial example, amounted to upwards of twenty-one thousand.

The environs of Vichy are said by several writers to be flat and uninteresting, but this observation, which is evidently copied from a well known Guide Book,

* "Guide Pratique Des Malades Aux Eaux de Vichy"—Par F. Barthez, M.D. 7th Edition. Paris, 1865.

does not apply, except to some parts of the immediate vicinity of the town. For the neighbouring country, at a short distance from Vichy, presents some of the most beautiful scenery and most interesting excursions in France. The river Allier, on which Vichy is situated, is a wide bed of dry gravel during the summer, and the newly finished suspension bridge, of immense length, which crosses the river, judging by the vibrations whenever it is crossed by a horse or vehicle, threatens before long to add to the debris in the stony valley which it spans.

The grand thermal establishment, situated at the extremity of the park, which it faces, is a handsome edifice, commenced in 1784, and lately re-modelled. Before the opening of the new and magnificent Casino the thermal establishment contained all the conveniences usually found in such establishments—concert hall, conversation and reading-rooms. There are about a hundred and fifty bath rooms in this building, of which eight are Douche. They are well constructed and clean. The baths are divided by the centre hall into equal parts, those on the left of the hall being for ladies, and those on the opposite side being used for gentlemen. Close to this is the new establishment, built in 1858, which also contains a hundred and fifty baths, twenty being Douche. In the former establishment the average charge for a mineral water bath is three francs; while in the latter it is only two francs.

The geological strata from which the springs of Vichy arise are the tertiary limestone and coal formations. There is reason to suppose that the plain of Vichy from Cusset to Gannet was, at one period, and for a long time, submerged, and formed a vast freshwater lake. The foregoing opinion is founded on the fossil remains of fresh water fish and aquatic plants imbedded in the limestone. This limestone is covered with sand, on which rests a rich alluvial soil, and under the rock the basin of mineral water is supposed by geologists to extend over an area of six miles.

There are nine principal mineral springs in Vichy, which are all alkaline, ferruginous, and highly charged with carbonic acid gas, and in all, minute traces of arsenic have been discovered. They are divided into cold and thermal, and also into natural and artesian. The former are in general warmer and less gaseous than the latter, which contain most iron. In all the springs the principal mineral ingredient is bi-carbonate of soda.

I shall first give the analysis of the principal springs, then make some remarks on each, and conclude with some general observations on the Vichy waters, and their medical use.

ANALYSIS OF THE PRINCIPAL MINERAL SOURCES OF VICHY.—(M. MOSSIER).

| Substances in one pound of water. | Grande Grille (grains). | Puits Carré. | Petit Puits Carré | Le Hôpital. | Boulet. | Celestins. | Lucas. |
|---|---|---|---|---|---|---|---|
| Carbonic acid ... | 6·81 | — | — | — | — |  | — |
| Carbonate of lime | 1·61 | 1·70 | 2·30 | 2·45 | 3·29 | Analysis imperfect. | 3·41 |
| Carbonate of magnesia ... | 0·30 | 0·30 | 0·30 | 0·27 | 0·35 |  | 0·41 |
| Carbonate of iron | 0·08 | 0·15 | — | 0·36 | 0·35 |  | 0·17 |
| Carbonate of soda | 34·61 | 32·40 | 36·30 | 33·52 | 42·70 | 32·07 | 28 56 |
| Sulphate of soda | 5·57 | 6·91 | 7·05 | 6·27 | 3 04 | 5·46 | 6·49 |
| Chloride of sodium ... | 3·15 | 3·88 | 2·64 | 1·10 | 0·47 | 3·45 | 7·31 |
| Total ... | 52·13 | 45·34 | 48·59 | 43·37 | 50·20 | 40·98 | 46· 5 |

Under the gallery behind the thermal establishment are found four of the mineral springs. The principal of these is the Grande-Grille, so named from the iron railing that surrounded it, which lies to the left extremity on entering this peristyle. The water rises in the centre of a very old-looking stone basin, thickly incrusted with a red ochreish deposit, by Sprudel-like, *persaltum*, jets, varying from two to twelve inches high, and has a temperature of one hundred and eight degrees. Its taste is rather disagreeable—saline, and somewhat ferruginous. It is employed internally and externally, and, as it keeps well, is largely exported. This source is principally used in gout, gastric complaints, dyspepsia, and affections of the liver.

The " Source des Mesdames " is at the opposite extremity of the same gallery as the last described spring. This source has only been employed medicinally within the last few years. It is contained within a separate enclosure from the other springs, and flows into a large basin lined with metal. Its temperature is only sixty-one degrees, and it is nearly identical in composition with the "Puits Lardy," containing a very large proportion of salts of iron, with traces of arsenic. As its name imports it is used in diseases peculiar to women, especially in anemia and chlorosis.

In the centre of the portico to the right of the entrance into the baths house is the " Source de Grande Puits Carrè," and opposite is the " Puits Chomel," or " Petit Puits." The former is inconveniently placed, and is now less used than the latter, with which it seems to have a common origin. The water from " Puits Chomel " is drawn up by a pump, and issues at the temperature of one hundred and seven degrees. Its taste is chalybeate, and slightly acidulous. Both these springs are prescribed in derangements of the stomach, and in some pulmonary cases, being then generally mixed with one third part of milk.

In the Place Rosalie is the " Hôpital Source," which is sheltered by a light iron pavilion, and is enclosed in a large stone basin, six or seven feet above the street. Within this, and rising a foot over it, is a smaller well, of great depth, through which the spring rises. Its taste is strongly saline, and by no means as sulphurous as might be supposed from its odour. The temperature is ninety-four degrees. The water rising in the shaft appeared to be in a state of violent ebullition, from the large volume of gas it contains; while that in the basin was covered by a greasy, whitish, flocculent scum. This is said to be the most aperient of the Vichy springs, and is prescribed to persons of a full habit, and soft leucophlegmatic constitution, suffering from gastric and intestinal derangement, especially chronic constipation, impaired digestion, and loss of appetite.

The " Célestins " are situated at the extremity of the old town, on the right bank of the river. The original spring, which has been lately deepened, and enclosed within a handsome pump room, contains more carbonic acid than any of the Vichy waters. It is the most exciting of them, and cannot be used with safety by persons of a delicate, nervous, sensitive temperament; nor by those suffering from diseases attended by inflammatory or hæmorrhagic symptoms. Very recently a case illustrative of this opinion came under my notice, in which a gentleman labouring under gastric derangement resulting from the gouty diathesis on his first arrival in Vichy, incautiously tried the Célestins, and, although he took but *one* dose of the water, suffered severely in consequence. Its use is principally restricted to renal and vesical complaints, and its action is strongly diuretic.

To the left of the last spring is the new source of the same name, which was discovered in 1858, and comes from an artificial rocky grotto under a handsome salle. In all respects it resembles the neighbouring spring. Within the same inclosure is a third source, the artesian well "de Lardy," a very strongly chalybeate, gaseous, and alkaline water, which contains a large proportion of sulphuretted hydrogen. Like the "Célestins" this is a very stimulant water, exciting the nervous system, and in some constitutions producing cerebral effects very like those of champagne, caused of course by the large amount of carbonic acid gas it contains. It is adapted to cases of anæmia and chlorosis, general weakness, and impoverishment of the blood, in persons of a soft lymphatic constitution.

The Vichy waters are alkaline, aperient, and alterative. Their most marked action is in chronic diseases of the stomach, and abdominal viscera, in indigestion and dyspepsia of almost every kind, in chronic hepatic disease, whether resulting from biliary calculi, fatty degeneration, or cirrhosis, and in hæmorrhoidal affections, which are so often connected with congestion of the liver. They are equally serviceable in enlargements of the spleen, especially when resulting from the intermittent fever of tropical climates. The aperient action of these springs renders them valuable in the treatment of chronic and obstinate constipation; and the same properties also point them out as desirable remedies for some cases of hypochondriasis and groundless depression of spirits. Vichy is a Spa also adapted for some cases of diseases of women, complicated with disordered menstruation, and for the anomalous "critical" complaints which often set in at the period of life when this function ceases.

Among the diseases for which this Spa is prescribed, renal affections hold an important place, and it is especially serviceable in chronic catarrh of the bladder and lithic or uric acid gravel.

Diabetes, which is a much more common affection than is generally supposed, brings each season a considerable number of invalids to Vichy. And if this Spa really possesses one-tenth of the curative properties some ascribe to it in these cases, it should justly render it one of, if not the most important Spa in Europe. For in so intractable a disease, over which physic has so little influence, any means which holds out even a rational hope of benefit, is no small boon. That Vichy does so, seems pretty well authenticated, not only by the statements of the local practitioners, but also by the evidence of patients who assert that they have been cured, or relieved here of diabetes, and this statement is corroborated by the experience of Parisian and French provincial physicians, who often advise their diabetic clients to try the Vichy springs. How the mineral water acts in such cases is a very obscure question; the most probable theory is that of Dr. Constantin James, who supposes it acts through the hepatic system,* and this idea seems in accordance with the teaching of modern physiologists, who ascribe the abnormal formation of saccharine matter primarily to derangement of the functions of the liver, by correcting which the Vichy waters are said to operate in these cases.

The complaint for which nine-tenths of the English visitors to Vichy drink these springs is gout, and I believe that the disappointment which so often drives these patients home again to patience and flannel, is the result of the misconceived ideas which prevail on this subject. It should be distinctly understood that Vichy water is not a specific, properly speaking, for the gout. It can only act indirectly on the gouty diathesis, by improving the tone of the digestive organs, augmenting the secretions and increasing the appetite, and thus the condition of the circulating fluids is amended. This effect is also aided by the alkaline salts of the Vichy water being taken into the system, and counteracting the abnormal acid principles peculiar to gout. The practical application of these observations is, that gouty patients who come to Vichy in quest of health should not consider that by the diligent use of the Spa alone for so many weeks they may expect much benefit, if they neglect the rules of living which I have pointed out in a former chapter. Gouty patients, whose disease springs from dietetic errors and neglect of exercise, come to Vichy; their appetite is increased

* Guide Pratique Aux Eux Minerals. p. 163. Paris, 1855.

by the change of air and foreign cookery, they indulge that factitious appetite fully at the table d'hôte, and then return home wondering why they ever went so far for so little good. The remedies for gout are, abstemiousness and exercise. Vichy water may aid, and aid materially, but it cannot supersede these.

Of the various varieties of gouty disease, that known as tonic, regular gout, by which term is meant gout attacking the articulations, especially the great toe, and occurring in robust plethoric persons, is that most suited for Vichy water. In such cases, the springs which are richest in bi-carbonate of soda—the " Celestins," for instance—are administered in the intervals of the complaint, and generally have the effect of rendering the attacks less severe, and also of lengthening the intervals between the gouty seizures.

Two miles from Vichy is the very ancient little town of CUSSET, situated between the rivulets Sichon and Jolan, and surrounded on all sides but the west by the mountains of Forez. The narrow winding streets, quaint old houses, and handsomely wooded Boulevards, all evince the antiquity of this place. Entering it from Vichy, we passed a large, stumpy-looking, round tower, now used as a prison, whose walls, twenty feet thick, resisted many a fierce attack from the Lords of Auvergne and Bourbonnais in the troublous days of Louis XI. Now, however, Cusset is an unimportant market town, remarkable only for its mineral waters. But even these seem to have attracted less attention than they deserve, and I am not aware of any published account of the springs of Cusset, with the exception of a short allusion to them in Dr. Aldridge's " German Spas and Vichy."

The mineral wells of Cusset appear to be all artesian, except the " Source Abattoir." The principal of these sources which I examined were—the springs in the " Course Napoleon," the " Elizabeth," and " St. Marie de Cusset," which rise within the same garden, and the latter of which supplies the thermal establishment, and the " Source Abattoir," in the yard of the public slaughter-house.

These springs belong to the same class of mineral waters as those of Vichy, than which they are, however, stronger, containing more carbonic acid gas, bi-carbonate of soda, and iron, than any of the Vichy sources. The " Elizabeth" well, for instance, is said by its proprietors to contain six times as much soda as any of the Vichy springs. The " St. Marie" is also very rich in the same salts. It supplies the bath house, which is a handsome building with reading saloons, pump rooms, and about thirty very neat and well-constructed douche and reclining baths.

The waters of Cusset, which are all cold, are used in the same class of cases as those of Vichy, but as they are said to be stronger, require still more caution in their administration.

In the same ancient province of Auvergne, though not in the same department as Vichy, there are some other mineral springs, less known than Vichy to English valetudinarians, but esteemed of considerable remedial power by French physicians and their patients. For the materials of the following brief notice of these Spas I am largely indebted to the assistance of my father, Dr. R. R. Madden, who has visited the springs of Auvergne this year since I have been in Vichy.

The Spas to which I would now invite attention are the thermal waters of Mont Dore, St. Nectaire, and Royat. The first of these, MONT DORE, may be reached from Vichy by railway to Clermont, and thence by coach. It lies about thirty miles from Clermont, in a small valley three thousand five hundred feet above the sea, and immediately under the Pic du Sancy, the highest mountain in central France. Between Clermont and Mont Dore the road which passes the remarkable mountain of the Puy de Dôme crosses the most singular, and to a geologist most interesting, volcanic district in Europe; on every side may be seen extinct craters, masses of the scoriæ and lava ejected from these craters, and vast blocks of basaltic rock, evidently of volcanic nature, all of which attest that this region was at one time the scene of convulsive, igneous action of incalculable force and activity

MONT DORE LES BAINS, although said to be one of the most ancient watering places in Europe, is but a mere village, containing several good hotels. There are eight mineral sources here, the temperature of which vary from 115° to 59°. The chief chemical ingredients in all these springs are bi-carbonate of potash, carbonate of lime, and sulphate of soda. Besides these, recent chemists have proved that the waters contain rather more than one millegramme of arsenic, in the form of arsenite of soda, in each litre of the water. Montaigne, who passed some time at this Spa, ascribes the good effects he witnessed and experienced here to the combined influence of change of climate, of scene, and of thought with exercise, without any reference to the waters; while the chemist attributes the same effects solely to the action of the arsenite of soda and other salts. For my own part, I am inclined to think that, as in most other debated points, there is some truth on both sides, and that the Sieur Michel de Montaigne had at least as much reason for his opinion as the modern professor; the cures which take place at Spas resulting from the combination of moral with physical agencies.

The cases in which the baths and waters of Mont Dore are prescribed are certain forms of chronic bronchitis, asthma, and laryngeal complaints, gastro-enteric, and uterine disorders marked by congestion, similar cases in which the liver is implicated, nervous maladies, such as neuralgia and sciatica, and scrofulous diseases, especially of children.

About fifteen miles from Mont Dore, near Murol, is the watering place of St. NECTAIRE, also in a volcanic district. There are seven thermal springs in this locality, the temperature of which varies between 75° and 110°. They are all akaline, ferruginous, and stimulant. They are principally used in cases of renal and hepatic disease, and especially in enlargements of the liver or spleen; they are also employed in ammenorrhœa, leucorrhœa, and gout.

The last of the watering places of central France to which I shall allude is ROYAT, which is eight miles from Clermont. The waters of Royat closely resemble those of Mont Dore, than which they are, however, one-third stronger. Royat is a Spa in considerable repute with many French physicians in the treatment of cases of scrofula, gout, and rheumatism.

## CHAPTER XXIII.

A VISIT TO THE SPAS OF THE PYRENEES.

Reminiscences of a walking tour in the High Pyrenees—The Pleasures of Hardship—General observations on the hot Springs of these Mountains—From Tarbes to Bagnères—The Bearnèse peasantry—BAGNERES-DE-BIGORRE: its waters, their History, and Action—Our Walk resumed —The Road to Lourdes—Valley of Argelez—The Poetic versus the Prosaic—Arrival in CAUTERETS—Accommodation and Resources—The Mineral Springs examined—Diseases in which they are prescribed.

THE localities described in the following chapter are, I believe, less familiar to British valetudinarian tourists than any part of Europe frequented by travellers in pursuit of health; and yet no country is richer in mineral waters of undoubted remedial potency than the French Pyrenees.

Unlike the Spas of Germany and Switzerland, crowded with travellers in quest of amusements, as well as with pilgrims to the fountains of health, the mineral springs of the Pyrenees are seldon visited by any who are not real invalids, and of these comparatively few are English valetudinarians. Therefore, I think, that the following reminiscences of an extended walking tour through the High Pyrenees, undertaken for the special purpose of visiting and examining the various Spas scattered throughout this vast range of mountain, may be found of some interest, inasmuch as they contain original information on a still unhackneyed subject.

The attractions of the Pyrenees are not, however, confined to the invalid traveller, but even for the pleasure tourist offer inducements for a pedestrian excursion in some respects superior to any in Switzerland. And for a man in health, what mode of travel affords hereafter such pabulum for memory, such a variety of incidents and such pleasant recollections as a pedestrian journey with a genial companion, and in fair weather, through so beautiful a country as the High Pyrenees? He who would attempt this, however, must be prepared to " rough it;" to endure fatigue, occasional inclemency of weather, meagre diet, and indifferent lodging, if he would go beyond the mere beaten track of tourists, and to dispense for the time with the luxuries, contenting himself with the necessaries of life.

It is strange how soon one gets accustomed to the hardships of this mode of life. Thus for instance, although previously unused to pedestrianism, within a few days after I commenced this walking tour, I found myself so braced up by the pure mountain atmosphere that I could walk without any material inconvenience from early morning till evening. During the journey we usually started on our day's walk before breakfast, and seldom halted much before noon, when we stopped at some wayside cottage, or shepherd's hut on the mountain, where hard black bread, with goat's cheese, was the only procurable refreshment. After a short rest we then resumed our way, often through snow or rain, till dusk, when we contrived to reach a hamlet where we might shelter; and considered ourselves fortunate if our hostess, who usually knew as much French as we did of the Bearnèse patois, could furnish our dinner with an omelet.

Such a programme may not look inviting, but this I know, that if the object of an autumn tour abroad be the improvement of health, impaired by attention to some absorbing pursuit, and a sedentary civic life, a pedestrian journey of this kind will do such a traveller more physical, as well as moral, good than would the same time spent in Paris, or in the gambling saloons of Baden-Baden. And even for travellers who are unable, or unwilling, to encounter such hardships and fatigue, the Pyrenean Spas offer resources. The railroad from Bordeaux now runs to the very centre of these mountains, and every Spa is within an easy drive from the train.

The scenery of the High Pyrenees far surpasses anything in Switzerland. Not like the bleak and desolate aspect of the Swiss mountains, those through which our pilgrimage to the Spas now leads us are, on their lower slopes, covered with green foliage and luxuriant meadows; above they are clad with forests of oak; and still higher, beyond these, lie the perpetual snows that supply the torrents which irrigate the rich plains below. This chain of mountain, forming a barricade between the Iberian and Gallic races, and extending close on three hundred miles from the Bay of Biscay to the Mediterranean, is cut into by numerous gorges and valleys of surpassing beauty, through some of which our route now passes.

In no part of Europe will the valetudinarian find so wide a choice of mineral and thermal springs to select from, within the same extent of country as in the Pyrenees, where some two hundred of these fountains of health have been discovered.

The mineral waters of the Pyrenees may be divided into three classes, viz. :—Sulphurous, Saline, and Ferruginous, and two-thirds of these springs belong to the first-named class, of which Cauterets, Barèges, Bagnères-de-Luchon, and Saint Sauveur are examples. The saline waters are illustrated by Bagnères-de-Bigorre and Dax; and the ferruginous by Castera-Verduzan, and Casteljaloux.

Our pedestrian journey through the Pyrenees commenced from Tarbes. We might, indeed, have gone by rail as far as Bagnères-de-Bigorre; but missing the morning train, found we should have had some hours to wait; instead of doing which, we started off on foot. We had no "impedimenta" but a light valise, which we carried by turns ; but still so hot was the weather, although it was in the middle of October, that even this light weight was sensibly felt.

On leaving the town we entered the wide valley of Tarbes, and enjoyed a magnificent view of the distant Pyrenees, in outline very like that of the Alps, as seen from Geneva, excepting, that instead of the dazzling whiteness of the Swiss mountains, these are dark and sombre. For the first six miles the road passes through a highly cultivated plain, where the peasantry were collecting the late harvest of maize, in long waggons, drawn by oxen. Every patch of ground is tilled, and the fields, irrigated from canals on each side of the roadway by rivulets through each farm, reminded me of the Moorish system of agriculture, still practised in the Vega of Granada.

About ten kilometres from Tarbes, the ascent of the Pyrenees commences. After a brief pause in the hamlet of Mont-Gaillard, where the hostess of the village inn prepared an omelet worthy of the Maison Dorèe, leaving the tilled plain behind, we entered into a hilly pasture country not less populous than the lowlands, and at nightfall arrived at Bagnères.

BAGNÈRES-DE-BIGORRE is an ancient town of nine thousand inhabitants, situated at the foot of the mountains, between the valleys of Tarbes and Campan; and may now be reached from Paris by railway, viâ Bordeaux, in thirty-six hours. It differs from most of the Pyrenean Spas, in being an important looking town, and not merely a few hotels and lodging-houses built around some mineral springs.

The aspect of Bagnères-de-Bigorre is very Spanish. The promenade is more like an Andalusian "Alameda" than a French "Boulevard;" the narrow winding streets, the projecting roofs of the houses, and the dress of the peasantry, the women wearing a head dress like the mantilla, and the men a long cloak, all reminded me of a Spanish scene. The streets, however, are remarkably clean, little streams of water flowing through each of them.

The bath houses and thermal establishment are large handsome stone buildings, and the hotels are numerous and good.

Next to the springs, the country about it, is the great attraction of Bagnères. The surrounding plain, watered by the Adour, trends into deep valleys of wonderful fertility, embosomed between green hills, which rise in successively higher ranges around the plain, while in the centre of the landscape is the vast pyramidal form of the Pic-du-Midi.

BAGNÈRES-DE-BIGORRE belongs to the class of saline sulphurous waters. There are a great number of springs here, some twenty of which have separate names, and are supposed to have different effects; but this difference seems to me a purely imaginary one. Bagnères is probably built over a subterranean stream of warm mineral water, which springs up whenever an opening is made in it. And, consequently, the several wells differ only accidentally, according to the strata they may pass through between this subterraneous river and the surface. Besides the sulphate of lime which is characteristic of all these sources, with few exceptions, they all contain more or less carbonate of iron. The largest amount of ferruginous matter is found in the " Dauphin," " Roc-de-Lannes," " Source-des-Yeux," and " Saint-Roch " springs; and the largest amount of sulphate of lime is found in the " Source de la Reine," in which that salt forms more than two-thirds of the mineral constituents of the water.

Bagnères-de-Bigorre, like so many other watering-places, was known to those enthusiastic hydropathists, the Romans, and many of their monuments and votive tablets have been discovered here.

Nearly all these springs are stimulant and exciting, and most of them act as saline, ferruginous tonics. They are accordingly indicated in chlorotic and anæmic cases; in menorrhagia; in chronic mucous discharges from either the urinary or the pulmonary organs, when unaccompanied by any inflammatory action; in hæmorrhoids, when produced or attended by habitual constipation; in dyspepsia and loss of appetite; and in some forms of enlargement of the liver and spleen.

The waters of " La Reine " and of " Lasserre," in doses of from five to six glasses, are considered as mildly laxative. The less strongly mineralised sources —" Le Source Foulon," " Petit Baréges," " Salut," and "Saule,"—are asserted to exercise an opposite effect on the system, to the stronger springs, being regarded as producing a sedative and soothing action on the nervous and circulatory systems.

Having spent some days in Bagnères, and examined all the mineral springs, we now resumed our journey to Cauterets. We were obliged to retrace our steps towards Tarbes as far as Mont-Gaillard, but had not proceeded more than a couple of miles when a heavy shower compelled us to seek shelter in a neighbouring church, from which we had heard the sounds of music, and where, on entering, we found the population of the district congregated. Ere we had advanced little more than a mile further on our journey, the rain again came down in torrents, but as Lourdes, where we must sleep, was a considerable distance off, we pushed on, and had a proof of the natural kindness of the people, for a farmer, seeing us pass, sent after us to beg us to return and take shelter. At Mont Gaillard the road to Lourdes turns away to the left, from the "Route Impèriale," and ascends one of the steepest spurs of the Pyrenees, which took us a couple of hours to surmount. But we were well compensated for this fatigue by the view from the table-land above, which extended as far as the eye could reach, over the apparently interminable ranges of lofty mountains, between which we caught occasional glimpses into the rich valleys that lie sheltered amongst them. At this point we encountered a stream of waggons, farmers mounted and on foot, and droves of oxen, horses, and pigs, which continued in an almost unbroken line for miles. These we found were returning from the great annual fair of Lourdes. Night came on, and we could see no sign of being nearer to our destination. The fair returning folks had disappeared, and the few passers we hailed in the dark could give us no information, as they spoke nothing but the Béarnese patois, of which we were completely ignorant, and we came to the conclusion that we must have passed Lourdes in the dark. Now and then a light in the distance would tempt us to urge on our weary limbs, but on coming near we would find the ignis-fatuus was but the twinklinglantern of some approaching waggon. At last, when we had almost abandoned all hope of reaching Lourdes that night, a turn in the road suddenly brought us into the dirty, narrow, and unlighted streets of the town, still noisy with the half-tipsy loiterers after the fair.

Lourdes, though remarkable for little except its antiquity, and the fortress on the hill above the town, yet boasts an excellent hotel, where a good supper and comfortable rooms soon removed every vestige of our toilsome journey.

Next morning, starting betimes before breakfast, we entered a barren, rocky valley, and ascending high above the impetuous Gave, by a road not much unlike that which those who have crossed the Irish Channel will remember as the Scalp, near Dublin, with mountains on all sides, whose loose rocks seem ready to fall on the traveller below. Emerging from this desolate region, we crossed the Gave by a handsome bridge, and entered the valley of Argalèz. The change from the sterile scene we had just left, to the cultivated and romantic beauty of this, the loveliest valley of the Pyrenees, seems as magical as that which Prospero's wand conjured up to Miranda's vision.

The vale of Argalèz is guarded at its entrance by the ruins of Castel Laubon, under which the road winds. In the centre of the plain is Junaclos, and in the far distance are the snow-covered peaks of the High Pyrenees. The lowlands, divided by the rapid river, are covered with maize, or else form rich pasturages. The orchards are luxuriant, and the trees, especially the walnut, lime, and cherry, attain a remarkable size, and their abundant fruit strews the road.

Beautiful as was the scene, our increasing appetite began to remind us of less poetic thoughts, and we were becoming painfully sensible that we had not yet breakfasted, when we reached Argelèz, in the centre of the valley. There we recruited our strength by a tolerable *dejeûner*, and rested till the sun had passed its meridian. Meanwhile, we gazed with undiminished pleasure on the view from this central position, which surpasses description. The "Val d'Azun," opening into the mountain on the west, lay spread out beneath us; to the south was the lofty Pic-de-Soulom, behind which three lesser snow capped mountains were conspicuous by their altitude from the surrounding chain; and in the immediate vicinity of Argelèz, were the ruined outlines of eight ancient castles.

Leaving Argelèz, above the road, on a wooded hill, is the monastery of St. Savin, of old a famed shrine for pilgrims in quest of bodily as well as spiritual health. The valley now begins to contract and becomes narrow near the Castle of Beaucens, under which is a warm mineral spring, and a small bath house. We now arrived at the village of Pierrefitte, at the foot of a lofty hill, which abruptly closes the valley of Argelèz.

At this point two gorges open in the mountains, the one to the left leading to Barèges, the other on the right to Cauterets. Leaving the former, and taking the latter route, we turned towards Cauterets, and commenced the long ascent before us by an ingeniously engineered road, leading up the side of the mountain in zigzags, supported over the ravine on buttresses of solid masonry, and in other places cut through the projecting rocks.

For miles the road passes through a most picturesque, though stern and wild country. The valley is but a dark and narrow gorge between two mountains. In places detached masses of limestone and marble of great size, appropriately named "Les Belles Horreurs," overhang the road. After some miles, the road brought us near the river side, where the Gave falls over a marble dam, which stretches across the gorge, and forms the "Côte du Limaçon," one of the many waterfalls of this torrent. Above this point the valley expands into a triangular form, the base of which is formed by an immense pyramidal mountain, at the foot of which lies Cauterets, where we arrived in time for dinner, after a walk of ten hours.

CAUTERETS is situated 3,000 feet above the sea, in a kind of *cul de sac*, formed by the high mountains, in which the French Route Impèriale ends, the only road beyond the village being by the footpaths which lead into the recesses of the Pyrenees, or into Spain. Above the town, on all sides, rise the precipitous mountains containing the mineral springs, and intersecting the narrow gorge of Cauterets, the Gave expands, and becomes still more rapid than in any other part of its course.

The houses of Cauterets are for the most part built of marble from the surrounding mountains, and form a long narrow street divided by an irregular square, containing the "Hotel de Paris" and some others, from which a shorter street leads up the side of the mountain to the thermal establishment. There are a couple of hundred houses, most of which are inhabited only during the season. The climate of Cauterets seemed to me damp and relaxing, while at times the wind rushes with great force through this gorge from the mountain passes.

The view from the hotel in the square just mentioned is very picturesque. On all sides rise the sharply cut outlines of the mountains above the village, covered with snow to within a quarter of a mile of the houses. To the left, looking from the balcony, is the bridge crossing the Gave, which, tumbling over the gigantic boulders in its bed, forms a series of miniature waterfalls, over which the stream rushes, and may be traced by a long line of white foam. On the opposite side, terminating the street leading up the hill, is the thermal establishment, a large and handsome building, approached by a long flight of steps leading into an open portico, supported on marble pillars. This bath house is one of the best in the Pyrenees, and is fitted up with every modern improvement in the baths. It is supplied from the sources " Des Espagnols " and " De César," and from the old bath house higher on the hill, which are conducted down by wooden pipes in the same way as those of Pfeffers are brought to Ragatz.

Upon the whole, Cauterets struck me as a very dull and inaccessible Spa, for any one who does not really require its mineral waters, to visit. The railway via Bagnères-de-Bigorre, has, however, now brought this watering place within a couple of days' journey from Paris. Moreover the beauty of the surrounding scenery, and the excursions which may be made from hence, through the High Pyrenees and into Spain, especially that by the " Pont d'Espagne " to the wild and romantic Lac de Gaube, attract many visitors to Cauterets, who have no need of its mineral springs.

CAUTERETS possesses twelve mineral sources, which

are scattered about the village, and some are at a considerable distance from it. For convenience of description they are divided into two groups, viz.:—*les sources de l'est* and *les sources du midi*. They differ in temperature from 131 degrees to 86 degrees, and present an important variety of composition, which, while it adds to the resources of the resident practitioner, renders it necessary for the patient to consult one of the local physicians before commencing his course of the waters. They all agree, however, in being rich in sulphuretted hydrogen gas, in silica, and in glairine, and in the rapidity with which the sulphates they contain are decomposed and changed into sulphites and hyposulphites.

LA RAILLÈRE, the most celebrated of the mineral springs of Cauterets, is about half-an-hour's walk from the town, on the road to the Lac de Gaube. The water is clear, and its temperature is 102 degrees; it is unctious to the touch, with a slightly sulphurous smell, and sweetish, mawkish taste. It is conducted into a commodious bath house, the douches in which are particularly well constructed. This spring is used in the same class of cases as those in which the Eaux-Bonnes are employed, namely, in chronic catarrhal affections of the respiratory organs, and in incipient phthisis. My own opinion of its value in tubercular disease of the lungs differs, however, from some of those who prescribe this remedy in such cases, and I think that more harm than good would generally result from sending a consumptive patient to Cauterets.

I was informed that it was a matter of traditional observation, long before Cauterets became so famous a watering-place as it now is, that the horses and cattle of the neighbourhood, when suffering from cold and other diseases, used, of their own accord, to resort to the then unguarded thermal springs, and by drinking the waters were cured.

The sources of "César" and "les Espagnols," which supply the grand bath house, are the most stimulating waters of Cauterets. Internally, they are used in chronic catarrh, and in some forms of asthma, and, as baths, they are ordered in cases of chronic rheumatism, certain skin diseases, and scrofulous affections.

"Les deux Pouces" are similar in action, but milder than the last described springs. The "Petit St. Sauveur" is a more soothing water, and is chiefly used in baths and douches, in leucorrhœa and some nervous affections. "Le-Pré" is a fountain administered in chronic rheumatism, and so also are the springs of "Mahourat" and "Le Bois," which are nearly a mile from the town.

## CHAPTER XXIV.

### THE FRENCH WATERING PLACES CONTINUED.

Journey to Lux—BARÈGES—The springs—ST. SAUVEUR - Account of this Spa—BAGNÈRES-DE-LUCHON, and its mineral sources—AMÉLIE-LES-BAINS—A long walk—Midnight in the High Pyrenees—Arrens—A passage under difficulties—Arrival in EAUX-BONNES—The waters and their medicinal uses—EAUX CHAUDES—The sources—Termination of our walking tour—PAU—The Chalybeate water of the Parc—DAX—"La Fontaine Chaude" Its history and present employment—Return to Paris—The springs of PASSY, AUTEUIL, and ENGHIEN-LES-BAINS.

FROM Cauterets we returned to Pierrefitte, where the valley of Argelèz terminates in two mountain passes; that on the right leading to the western Pyrenees, and that on the left to the eastern Spas of these mountains. Our present road lies through the latter. Traversing the bridge of Ville-Longue we enter the ravine leading to Lux. The sides of the valley, as we advance, soon close in, and the rugged, pointed rocks, on either side, are only separated by a deep but narrow fissure, through which flows the Gave. For, like almost every Pyrenean pass, this has its *gave*, or torrent, high above which the causeway is carried, on a mere ledge of rock, and where no ledge exists, on a series of marble arches.

Before reaching the town this gorge expands into a valley, on the left of which Lux is situated, on the extremity of a small plateaux, and immediately under the ruins of an ancient castle At Lux the road to Barèges branches off to the left, and after a steep ascent of nearly four miles, enters the dark and cheerless valley of the Bastan, in which is situated this very important watering place.

BARÈGES is a small village, consisting of one long street, of about a hundred houses, built of stone, and standing immediately over the torrent, and is as wild and desolate a place as can well be imagined.

Being the most elevated watering place in Europe, the climate, even in summer, is cold and variable, and in winter it is such as to render the village uninhabitable. None but those who absolutely require the waters are to be met with in Barèges; for nothing else could, I think, induce anyone to pass a single week in this village. And yet, however, between six and seven thousand invalids reside here during the short season. Great, therefore, must be the medical virtues of the springs, which can thus attract the votaries of fashion, and of pleasure, to so remote, inaccessible, and dreary an abode as this.

There are nine mineral springs in Barèges, the temperature of which vary from 86 degrees to 112 degrees. Eight of these sources are contained within the bathing establishment, in the centre of the town. In chemical composition these waters differ little from each other, being all examples of what French writers describe as, " les eaux sulfuréuses sodiques."* They are perfectly clear, have a strong and nauseous sulphurous taste, and well marked "hepatic" odour. They are more fixed, and change less by keeping than the other Pyrenean sulphurous waters. The sulphuretted hydrogen gas escaped from them with remarkable slowness, and there is no white sulphurous precipitate from these, as there generally is from other sulphurous waters; but they all contain a considerable amount of the peculiar, pseudo-organic, unctuous substance called " Barègine."

The saline matters found in the nine sources of Barèges are nearly identical. They principally consist of the sulphates of soda and lime, silicates of the same salts, and chloride of sodium, with traces of oxide of iron and iodine.

The primary action of the Barèges Spa is stimulant and tonic, producing considerable nervous and vascular excitement; and accordingly it is best suited for persons of lymphatic and scrofulous diathesis, and it should be especially avoided by those of a plethoric habit of body, by pulmonary invalids, and also by valetudinarians suffering from any hæmorrhagic or congestive disease.

In the form of baths, these waters are applicable in the treatment of cases of old wounds, either breaking out

* M. M. Petréquin et Socquet—"Traite Général des Eaux minérales,' p. 413. Lyon, 1859.

afresh, or causing recurring pain, in carious disease of the bones, in articular complaints, whether rheumatic or resulting from injury. They are largely used in obstinate skin diseases, more especially when of secondary, or of mercurial origin; also in chronic rheumatism and scrofulous affections. The duration of the course must be regulated by the effects of the waters. The full time is six weeks; but in many cases half that period might be dangerous to the patient's life.

About a mile from Lux is a Spa in every respect different from that which has been last described, viz., SAINT-SAUVEUR. The situation of this village, at the entrance of the plain of Gavarnie, within reach of the most sublime scenery in France, *i.e.*, the Brèche de Roland, Mont Perdu, and the Vignemale, is most attractive. The watering place itself is a mere hamlet of thirty or forty houses, connected with Lux by an avenue of poplars. There are a couple of good inns, besides which the accommodation of the lodging-houses is comfortable, the charges are moderate, and the thermal establishment is very complete.

There is but one mineral source in Saint Sauveur; although there are four or five outlets or springs from this, which some writers mistake for distinct sources, and describe as such.

The water is saline and alkaline. It is perfectly limpid, is warm, 95 degrees, and has a soapy or unctuous taste, caused by the amount of glairine suspended in it.

The waters of St. Sauveur are used as baths, injections, and for drinking. Their medicinal action seems due to their temperature, the amount of nitrogen and glairine they contain, and their alkalinity. In their influence on disease they somewhat resemble the springs of Wildbad, and exert a peculiarly sedative influence on persons of a nervous and irritable temperament. They are prescribed in hysterical cases, in dyspepsia, in vesical catarrh, in uterine affections, and leucorrhœa, and in neuralgia and sciatica.

The next of the Pyrenean Spas, in the order of their position, is BAGNÈRES-DE-LUCHON, which lies close to the Spanish frontier in a valley of the Pyrenees, and the department of the Upper Garronne, and may now be reached in about forty hours from Paris by the railway to Bagnères-de-Bigorre, and thence by diligence. The situation of Luchon, at the foot of the highest range of the Pyrenees, six miles from the Spanish frontier, in a cultivated valley, is very pretty. The entrance to the town from the north is through a a long avenue of limes, which is intersected by a handsome promenade, and leads into the " Cours d'Etigny." This is a kind of straggling street, half a mile long, lined with hotels and lodging houses, and terminating in the bath house. The old part of the town consists of a few narrow lanes behind this. The resident population amounts to about three thousand, to which is to be added eight or nine thousand visitors in the time of the season.

The accommodation for invalid visitors at Luchon is better than at most of the Pyrenean Spas; there are nine or ten very tolerable hotels, and a fair choice of apartments. Living is comparatively very cheap and good, game being abundant, and the fish from the mountain streams is excellent. Not only are the comforts, but the luxuries of civilised life, such as balls, concerts, clubs, &c., attainable in Luchon, and in this respect it comes nearer to the German watering places than the other Pyrenean health resorts. Besides this picturesque situation of the town, and its proximity to the Spanish frontier, all combine to attract valetudinarians to Luchon.

The bath house is one of the largest in the Pyrenees, and is formed by eight pavilions, in which are found every form and variety of bath, fitted up with all the modern improvements. There are about thirty distinct mineral springs in Luchon, the temperature of which varies from 150 degrees to 60 degrees; but differ very little in other respects.

At the fountains the water has a strongly sulphurous smell, but appears limpid; shortly after it is taken from the source, however, it becomes milky and cloudy. This change is ascribed to the partial decomposition produced by an excess of silicates and silicic acid. The springs are all sulphurous and saline; their principal saline contents are the sulphate and hyposulphate of soda, with silicate of soda.

The water is used principally for bathing, but is also

taken internally, in doses of from two to three small glasses, either pure or with an equal part of milk. The diseases in which Luchon is resorted to are—cutaneous affections, such as eczema, syphilitic eruptions, lichen and impetigo. The springs are also employed in chronic ulcers, chronic rheumatism, and passive glandular enlargements, in articular diseases, and caries. They are also strongly recommended in many cases of scrofulous enlargements, hypochondriasis, and dyspepsia.

Amélie-les Bains, in the department of the Eastern Pyrenees, on the route from Perpignan to Barcelona, is now one of the most rising health-resorts in the south of France, being frequented by valetudinarians not only in summer, on account of its mineral waters, but also in winter by pulmonary invalids, though in the latter case, in my opinion, often with somewhat questionable judgment, on account of the mildness of the climate.

The village, which contains about five hundred inhabitants, is built in semi-circular form, at the foot of a hill on the right bank of the river Tech, being situated in a pretty valley, about three miles from Arles. It is evidently very ancient, and so far back as the seventh century is said to have been conferred by Charlemagne upon the Benedictine convent of Arles. The mineral waters of Amélie issue from numerous sources, which differ little from each other, being all warm, sulphurous, and very gaseous. There are three separate thermal establishments, which are principally employed in the treatment of chronic rheumatism and arthritic complaints, in skin diseases, in nervous maladies, especially neuralgia and sciatica, in affections of the kidneys and urinary organs, such as vesical catarrh, gravel, and calculus; in diseases peculiar to women, especially leucorrhœa and irregularities of the catamenia. Amélie is also frequented by scrofulous patients, and is said to be productive of great benefit in glandular and articular diseases of this class. It is, moreover, though I believe with much less utility, prescribed in chronic laryngeal and bronchial complaints, in dyspepsia, in abdominal and hepatic enlargements, in hypochondriasis, and even in some cases of consumption.

To reach the next Spa on our list we had to retrace our footsteps back to Pierrefitte, at the head of the valley of Argelez. Our journey thence in winter, when the mountain footpath we traversed was often indistinguishable in the fast falling snow, across a country perfectly unknown to either of us, more than once made us regret that we had disregarded the counsel of those who had endeavoured to disuade us from a walking tour through the Pyrenees at that season. At length, however, we reached the village of Arrens; but when we did so it was near midnight, and not an inhabitant did we meet, nor even a light could we see. Wet and weary as we were, the prospect of passing the night in the open air in a snow storm was not particularly cheering; so, advancing to the largest house in the village, I hammered at the door with a stone, for in this primitive spot knockers are unknown, till I aroused the inmates, and by them was conducted to the inn, where we finally obtained admittance.

Next day we pushed on shortly after daybreak. Our landlord advised us against attempting the Col de Tort, as he said a storm was impending, but, mistrusting the disinterestedness of his suggestion, we started. The first persons we encountered after an hour's walk, when we were ascending the mountain, were a couple of herds, driving their sheep into the valley, who repeated the same ill-bodings, and advised us to return to Arrens, as the path over the Col was covered with snow. However, as it looked clear and bright we pursued our way, and only when it was too late to turn back we found our error. For after a long, toilsome ascent, when we had crossed the first Col, and were passing the narrow footpath which overhangs the deep ravine under the Pic de Gabiscos, a hurricane suddenly arose, drifting snow and sleet, and large stones were hurled along as dust; we were forced back by the irresistible power of the wind; it was even impossible for us to stand, and it was only by throwing ourselves prostrate that we escaped being blown over the precipice. The wind came rushing down in squalls every few minutes, and between each of these violent gusts we had only time to advance a few perches before we were again forced down. At length, when nearly exhausted, we fortunately reached a wooden hut, occupied by a party of engineers, surveying a new *Route Impériale* across the Col.

There we procured some refreshment, and after a short rest crossed the Col de Tort, over which the path was undistinguishable under the accumulated snow, and finally arrived at Eaux-Bonnes, after a walk of twelve hours.

EAUX BONNES is a small village in the Basses-Pyrenees in the valley of Ossau, about three miles from the valley of Laruns, and twenty-five from Pau. Its situation, in a deep valley inclosed by lofty mountains, is highly picturesque. It consists of about fifty houses, for the most part large and handsome, built of marble, some five or six of which are hotels. The mountains approach so close to the village that they have been quarried in every direction to make way for the modern extension of the place.

The mineral sources are found at the foot of Mount Trésor; they all spring from a limestone formation, under which is granite rock. There are three springs, viz.: "La Vieille" (ou la Buvette), the temperature of which is 92 degrees; "La Nouvelle," temperature 88 degrees; and "La Source d'Eau bas," temperature 90 degrees. Besides these there are two other springs immediately outside the village; one of these wells, which is cold, is the only source here used internally.

The thermal waters of Eaux Bonnes are all very

gaseous, are limpid, and have a sulphurous smell and unctuous taste. They belong to the class of alkaline sulphurous waters; but are less alkaline, and contain less silica and more sulphate of lime than the other Pyrenean waters of the same class, excepting Eaux Chaudes only.

In their mode of action the springs of Eaux Bonnes resemble the neighbouring sulphurous Spas, their primary effects being stimulant, though less exciting than some of the other Pyrenean waters. They require, however, to be used with equal caution in small doses, commencing with a quarter of a tumbler full, which may be gradually increased, if it produces no ill symptoms, to two glassfuls a day.

The season nominally lasts from May to the end of September; but the weather generally becomes so cold during the latter month that few invalids could remain.

Formerly the principal use of the waters of Bonnes was externally in baths, in the treatment of old and painful wounds, ulcers, and other surgical diseases. Thus the name of "Arquebuscades," by which they were described by former writers, was derived from the wounded musketeers, who, after the battle of Pavia, were brought hither by Jean d'Albret, to be cured. Now, however, the chief employment of these springs is internally in the treatment of chronic pulmonary affections, among which is included phthisis. My own opinion of the inefficacy of mineral waters in the treatment of consumption has been expressed so often in the preceding pages that I need not here repeat it. But I may observe that the climate of Eaux Bonnes seems to me particularly unsuitable for consumptive patients.

The sulphurous springs of Bonnes are used in a great variety of complaints, and one writer of the last century goes so far as to say " Je ne connait pas des maladies auxquelles les Eaux des Bonnes ne puissent convenir," with the exception he, however, adds " of those in which there may be danger of quickening the circulation."[*] It is principally used in chronic maladies of the abdominal viscera, in hypochondriasis, and hysteria, in obstinate intermittent fevers, in chronic catarrhal affections, and in baths is employed in the treatment of chronic ulcers, fistula, and caries.

EAUX CHAUDES, in the Low Pyrenees, is situated in a wild and sombre mountain gorge in the valley of Ossau, on the right bank of the Gave, four miles from Eaux Bonnes. The road which leads from Laruns to Eaux Chaudes is in part cut out of the mountains, and in part is laid on an artifical platform along the torrent.

The approach to this watering place, through this dark and sombre ravine, is calculated to depress a nervous patient, and the aspect of the village itself, consisting of a few straggling houses, which stand on

---

[*] Théophile Bordeu—" Lettres sur les Eaux Minérales du Béarn," &c., 1746.

a narrow ledge between the mountains and the Gave, is far from cheerful. But still, though presenting nothing but its waters to attract visitors, Eaux Chaudes is frequented by a considerable number of invalid residents each season. There are four or five good hotels, and the expenses of living here are very moderate.

There are six mineral springs at this Spa, all of which issue at the junction of the granite and limestone formations at the foot of the mountain, which divides the valleys of Eaux Bonnes from that of Eaux Chaudes.

The thermal establishment is supplied by three sources—" Le Clôt," " Le Rey," and " l'Esquirette,"—which are all used both internally and as baths. Notwithstanding its name, none of the springs of Eaux Chaudes have an elevated temperature. Thus, the temperature of " Le Clôt," the warmest of the sources, is 97 degrees; " l'Esquirette," 90 degrees; " Le Rey," 93 degrees; " Baudot," 81 degrees; " Larressecq," 77 degrees; and one source is quite cold, namely, " Mainvielle," the temperature of which is 52 degrees. They are all but slightly mineralised, contain little sulphurous matter, but are rich in glairine. In appearance the water is clear, its taste is insipid, and its smell that of stale eggs, owing to the very large volume of sulphuretted hydrogen gas it contains. The principal saline contents of this Spa are chloride of sodium, sulphate of lime, silicate of lime, sulphate of soda, and carbonate of soda.

The mineral waters of Eaux Bonnes are all more or less stimulating and exciting, acting with equal energy on the nervous and circulating systems. They are, therefore, principally used in the treatment of obstinate chronic skin diseases, in scrofulous swellings, and articular enlargements, in chlorosis and amenorrhœa, in chronic rheumatism, sciatica and neuralgia, old wounds and ulcers, and in some cases of scrofulous ophthalmia. The following is an example of the occasional utility of these waters in certain forms of spinal disease:—A young gentleman had long suffered from spinal irritation, some of the lower dorsal vertebræ were very prominent; there was tenderness on pressure in the lumbar region, with loss of power in the extremities, and other symptoms indicative of compression of the spinal chord. Having been treated at home in the ordinary manner for a considerable time, without benefit, he was sent to Eaux Chaudes by Dr. Smith, of Pau, and at that watering place in one season, as I was assured, the protuberance and spinal tenderness disappeared, and he regained full power over his limbs. Another case also related to me by that same accomplished physician, the late Dr. Smith, to whom I was indebted for much most valuable information on this Spa, is a strong testimony to the efficacy of the Eaux Chaudes, in some instances of chronic rheumatism. An old gentleman, who, when I was last in Pau, still enjoyed comparative strength as well as health, although then upwards of ninety years of age, was met by Dr. Smith one day, fifteen years previously, feebly tottering along on crutches; the doctor, concluding that his old acquaintance was really "on his last legs," condoled with him on his state of health. "Ce n'est rien du tout, mon ami," replied the invalid, "je me vais aux Eaux Chaudes et au bout de quinze jours je serrai gai." A short time afterwards Dr. Smith met him at Eaux Chaudes, walking along as briskly as he had for years previously.

The dose of Eaux Chaudes water varies, of course, according to the condition of the patient and the source he uses: the warmer springs, being the most stimulating, require the utmost caution in their use, and it should be borne in mind that none of these sources can be employed with safety, excepting in accordance with the advice of a resident physician.

To understand this caution it is only necessary to glance at the physiological action of the Eaux Chaudes springs. I have already stated that they are stimulant, and excite the circulation as well as the nervous system. Thus, they produce a feeling of restlessness, amounting sometimes to absolute insomnia. They increase the secretions, especially from the skin and kidneys, and often manifest their action on the skin by occasioning a specific cutaneous eruption. I need now hardly add a warning to all invalids suffering from diseases in which an agent that, like this mineral water, has the property of disturbing and quickening the circulation, might prove injurious, to avoid Eaux Chaudes. No patient labouring under any tendency to inflammatory or hæmorrhagic disease, or of a plethoric habit of body, or nervous temperament, should use these waters. The following cases suffice to prove that this warning is neither unnecessary nor merely theoretical. A friend of mine, himself a medical man, came to reside for a season in Eaux Chaudes, and, as is the custom there, hired a swimming bath by the month, for a certain time each day. The ladies of the family used the bath early in the morning, and later in the day the gentlemen did so. The first day they used it, his wife suffered from a buzzing noise in the head, his eldest daughter was seized with palpitation as soon as she left the bath, and had to lie on a sofa all day; his youngest daughter slept little the following night; his eldest son had a transient feverishness, which he ascribed to this bath; and the only one of the family who did not complain soon after the bath was my friend's youngest son.

Even a very limited use of the Eaux Chaudes may, in some cases, if the patient be not in a fit state for this remedy, occasion unpleasant effects. The late Dr. Smith told me he once prescribed a foot-bath of the hot mineral water to a lady suffering from a catarrhal affection, and that this pedeluvium produced distressing symptoms of cardiac derangement and faintness.

Here ended my walking tour through the Pyrenees. From Laruns I drove to Pau, and renewed my acquaintance with the ancient capital of Bearn and Navarre. As, however, I have described this town very fully in my work " On Change of Climate," published three years ago, I deem it unnecessary to do so again in this place, especially as Pau, though celebrated as a sanatorium for certain classes of patients, is yet of very trifling importance as a watering place.

The mineral spring of Pau is situated immediately without the town, near the railway station, between the Park and the river. This source is not many years in use, as the inscription on it denotes :—" Cette source ferrugineuse à été découvert par le Sr. Bigot le 21re Mars, 1853, et Analysèe par M. le Docteur Fontan en Mai, 1854." The water is chalybeate, possesses a slight ferruginous taste, is cold and limpid, and leaves a ferruginous tinge in the stone basin, into which it is received. Its use, which is very limited, need not be here dwelt on, as it differs in no wise from what I have already spoken of in the introduction, as the general action of all simple chalybeate mineral waters.

Having rested ourselves after our long pedestrian journey through the mountains, by a short stay in Pau, we turned our thoughts homewards, and, leaving Pau by train in the afternoon, arrived the same evening in Dax. When I recollected my last departure three years previously from Pau, and the drive of eighteen hours thence across Gascony to Toulouse, in a small diligence, crowded to excess, and my previous journey by coach from Aire to Pau, before the railway had connected the capital of Bearn to the civilised world, I fully realized the advantages of modern civilization in this respect, at least.

The very ancient town of Dax on the Adour, since the completion of the railway from Bordeaux to Pau, is seldom visited by tourists, though one of the oldest watering places in Europe. So far back as the tenth century, the fountain of "Nelse" was frequented by patients from every part of Europe, but now, excepting the inhabitants of the department, hardly a single invalid is attracted to Dax by its thermal springs.

The hotel which we stayed at was in the suburb of St. Paul de Dax, which is nearer to the railway than the town, and next morning we proceeded to visit the springs. From this suburb, a short walk, crossing the Adour by a long bridge, brought us into the town, which lies on the left bank of the river. The aspect of the place is quaint and mediæval, enclosed within ancient fortified Gothic walls, containing numerous remains; a remarkable church, by Vauban; narrow, winding streets; and a population of about six thousand inhabitants.

Immediately after passing the bridge, which connects the suburb of St. Paul to the town of Dax, we came upon the principal thermal source, situated in a small, mean square, the "Place de la Fontaine Chaude." The water is contained within a remarkable structure—a large, oblong building, evidently of considerable antiquity, although apparently unfinished, being roofless. This edifice forms a reservoir of hot mineral water, some seventy feet from end to end, by fifty in width. From this vast body of thermal water a dense volume of steam arises, which is visible a considerable distance from Dax. The water is clear, and its temperature around the edge of the basin nowhere exceeds one hundred and twenty-five degrees, and at the outlet to which it is conveyed direct from the spring it only reaches one hundred and thirty degrees. This, however, is merely the temperature of the water, which covers this open space of three thousand, five hundred square feet, to the depth of two feet and-a-half only, and must, therefore, lose no small part of its caloric by the exposure. The temperature of the water in the shaft, through which the spring rises into the basin, is one hundred and fifty-six degrees. There is a tradition that Philip the Fifth, who passed through Dax on his way to assume the Spanish crown, ordered this well to be sounded, when a thousand "brasses" of cord being found insufficient for the purpose, the attempt was abandoned, and the shaft was pronounced unfathomable. This opinion was held by all the writers on the subject, until M. de Secondat published a memoire, in 1775,* in which he shewed that the depth did not exceed four toises, or twenty five feet. Submerged in the water, and adhering to the

* Dictionnaire Des Eaux Minéraels. Tome Premier, p. 291. Paris, 1775.

walls of the basin, I observed vast quantities of a minute aquatic plant, apparently one of the Algœ, consisting of a small cell. But so minute was it, that without a microscopic examination, it would have been impossible to pronounce whether these cells were of a vegetable or animal character. In either case, however, their existence and vitality in so hot a fluid as this is equally interesting.

The scene around the front of the basin when we first visited it in the early morning was very curious. The whole population of Dax were apparently assembled in this little square—every man, woman and child with a large, peculiarly-shaped tin vessel, strapped over their shoulders, and each patiently awaited their turn to fill these at one of the spouts by which the warm water issues from the basin. I remarked that before doing so, however, it was the invariable rule to take a deep draught of the thermal fluid. This water supplies most of the domestic and culinary purposes of the people of Dax, to whom it saves no small expense for fuel.

Besides this, there are two other thermal springs—"la Source des Fossés," and "des Bagnots;" one is more within the town than that last described, and a new thermal establishment was being constructed around it; the other, near the bridge, on the river side, is enclosed in an open pavilion under a glass roof; both, however, resemble the "Fontaine Chaude" very nearly.

The mineral springs of Dax are almost unknown in this country, and I am not acquainted with any English work that gives an account of these ancient *thermæ*, nor, indeed, have they fared much better in modern French books on mineral waters. They all belong to the class of mild saline thermal waters, their chief chemical ingredients being the sulphates of lime and soda, with a little common salt, and carbonate of magnesia. These salts of themselves could only produce a slight aperient action, and it must, therefore, be in their high temperature that we must seek the explanation of the active therapeutic effects of the thermal baths, and waters of Dax. They are commonly used, and with the happiest result, by the inhabitants of the Landes, in cases of chronic rheumatism, and rheumatic arthritis, causing impairment of the joints; and in contraction of the muscles, following recovery from severe surgical disease. They were prescribed by the French physicians of the last century, in certain forms of paralysis, and also in pulmonary affections, especially asthma. They are no longer used in these cases, and I think this disuse in paralytic and pulmonary complaints is only rational. But, externally employed, the baths of Dax are useful in many diseases in which a remedy that acts upon the skin, stimulating its capillaries, is indicated.

From Dax, the railway by Bordeaux and Orleans brings the traveller up to Paris in twenty-four hours. We, however, did not avail ourselves of this quick transit homewards, as we diverged from the direct route more than once to visit other places. But as none of these contain mineral springs, this chapter might close here did I not think a very short notice of the watering places in the neighbourhood of Paris might prove useful to many who would not care to journey so far as the Spas I have just described.

Within the recently extended limits of Paris, at PASSY, there exists a very strong ferruginous water—so strong that before it is used internally, it is necessary to allow it to stand exposed to the air for some time, until it deposits a ferruginous sediment which falls rapidly, being only suspended, not dissolved, owing to the want of sufficient carbonic acid gas. There are four sources, whose chemical composition is nearly identical. The chief mineral constituents of all these springs, (of which that called "la source Nouvelle" is the strongest,) are sulphates of iron, lime, magnesia, and soda. The Passy waters are used externally in the treatment of chronic ulcers, and in cases of leucorrhœa. They are also prescribed internally, with the precaution of allowing the water to deposit the suspended salts as I have just described, in doses of from one to six small glassfuls in cases of general and local anæmia, chlorosis, fluor albus, intermittent fevers, atonic dyspepsia, and diarrhœa, and, in a word, in all those diseases connected with poverty of blood, in which ferruginous tonics of this class are indicated.

In the neighbouring suburb of AUTEUIL a somewhat similar mineral water exists. This spring is known as the "source de Quicherat," and from the sixteenth century to the present time, has been resorted to by the Parisian bourgeois in the same class of cases as the wells of Passy are employed in.

The last of the French watering places which I visited was ENGHIEN, also in the neighbourhood of Paris, though not in the same department. I am indebted for my acquaintance with this Spa to a most hospitable Irish gentleman, Mr. B., residing in the adjoining village of Cormeilles, with whom I was then staying, and who kindly acted as my guide to this source.

For the last century French writers have been ecstatic in their praise when describing the scenery and situation of Enghien. Thus M. Reveille Parise, writing of Enghien, says—"Là se présente aux regards le plus magnifique, les plus gracieux, le plus attrayant des spectacles. Tout y charme, tout y retient, tout y séduit." And thus he continues to express his admiration of this place for whole pages. To me, however, Enghien appeared a rather prettily situated village, overlooking the valley of Montmorency. It is situated on a small lake, on the immediate borders of which are numbers of small, fantastically-shaped country villas, some of which are said to belong to persons of the *demi monde* society of Paris. The neighbouring hills shelter Enghien from northerly winds, but do not protect it from cold, easterly breezes. Enghien possesses a large thermal establishment open all the year round. This building faces the lake, and contains about a hundred bath rooms, besides excellent accommodation for invalid boarders, who can enjoy the use of the baths, and live very well in the establishment, or in the hotels connected with it, for about five pounds a week.

The facility of reaching Enghien from Paris is one of its chief advantages, as it may be approached in twenty-five minutes by either the northern or western railways from the city.

There are seven mineral springs at Enghien, all belonging to the class of cold sulphurous waters, and principally owing their medicinal properties to the very large quantity of sulphuretted hydrogen gas they contain. Their chief difference from the sulphurous waters of the Pyrenees is, that the wells of Enghien are cold, contain no barégine, nor salts of soda, their most important mineral constituent being sulphate of lime.

Of these sources only two, viz., " Du Roi " and " Des Eaux," are used for drinking, the others being too strong for that purpose. The Enghien waters are powerful stimulants. In small doses, conjoined with the baths, they increase the appetite, quicken the pulse, and excite a determination of blood to the skin. If taken for some days without intermission, or if the dose be at all too much, they occasion " Spa fever," sometimes of a most dangerous type.

These waters should, I think, be used with more caution than they generally are, for they appear to be employed in the most incautious manner, and in every class of chronic disease. I have already shown that strong mineral waters, like all remedies of any value, are two-edged weapons, not less powerful for evil than for good, and the sulphurous waters of Enghien are certainly no exception to this rule. They are especially liable to occasion cerebral congestion, and, therefore, I need hardly add they should never be used in any case in which determination of blood to the head is apprehended, nor when the heart or large vessels are engaged; in short, the waters of Enghien should never be prescribed in any case in which it might be injurious to stimulate the nervous system or to excite or quicken the circulation.

Small doses of the water should always be commenced with, and from half a glass to a small tumbler full is quite sufficient each morning. The baths should not be taken every day, nor should the patient remain many minutes in them. Both the baths and the waters should be discontinued if they produce any febrile irritation, headache, or cerebral excitement whatever; and the patient should not allow himself to be influenced by the advice of the managers of the thermal establishment if it be in opposition to these plain rules.

From the abuse, we now come to consider the use of this Spa, and we find that Enghien, like other sulphurous waters, is employed, and sometimes very efficaciously, in the treatment of chronic skin diseases, scrofulous affections, certain mucous discharges, some chronic glandular and articular enlargements, in some cases of chronic pains in the osseous structures, and other "secondary symptoms" as well as in chronic rheumatic affections.

## CHAPTER XXV.

### THE SPAS OF ITALY.

Preliminary remarks on the Italian mineral waters—The springs of Lombardy and Tuscany—AQUI, ABANO, Pisa and Lucca—General and medical account of these Spas—Monte Catino—The watering places of the Roman states—Civita-Vecchia, Viterbo, and Porretta described—Reminiscences of a summer in Naples—Escape to Castellamare—Advantages of this climate—The mineral springs, their analysis and therapeutic use—The island of Ischia—Its thermal sources and "Stufa"—Accommodation and resources for invalid visitors—The medical properties and mode of using these baths and waters—Conclusion.

FEW countries are so rich in mineral and thermal springs as Italy. Comparatively few of the Italian Spas however, are now resorted to by foreign invalids, and hence a very brief account of the most important of these mineral sources will here suffice.

Before giving my ideas of such of these fountains of Hygeia, as I have visited, I may premise that almost all the Italian Spas are situated in the valleys at the foot of the mountain chains, which intersect that country. Many years since Dr. Gairdner remarked, that " in the prolongation, southwards of the Italian peninsula, all its mineral waters are met with on the Mediterranean side of the Appenines, and none on the Adriatic."* The explanation of this curious fact is to be found, as I have shewn, in the introductory chapter of this work, in the position of the volcanoes, and other evidences of subterranean igneous action in that part of Italy.

Commencing our tour through the Italian watering places in Lombardy, the first of these Spas that we meet with is ACQUI, a very ancient town of nine thousand inhabitants, in a mountainous district, about thirty miles from Genoa, and a drive of an hour and a half by railway from Alessandria. The mineral springs originate in limestone rock, and are divided into cold and thermal sources. They are mildly sulphurous; but are so slightly charged with chemical ingredients, that, diluted with half their quantity of water, they are employed by the housewives of Acqui for all domestic and culinary purposes. The warmest source in the centre of the town has a temperature of 124 degrees; it is slightly sulphurous and saline, and is perfectly clear and limpid.

Acqui is seldom resorted to as a sulphurous Spa, being inferior in chemical strength to most waters of that class. But employed externally, mixed with the soil through which it rises in the form of mud-baths, the springs are remedies of considerable power. A large bath establishment has been constructed about a mile from the town, and here the "humus," or mineralised mud, is collected in small chambers, into which

* "Essay on Mineral and Thermal Springs," by Meredith Gairdner, M.D., p. 141. Edinburgh, 1832.

the patient enters, and lying down, is gradually and completely, with the exception of his head, covered with a thick layer of the " humus," as hot as he can bear it, and remains thus immersed in the semi-fluid black broth, about three-quarters of an hour; immediately after which a warm bath of the mineral water is administered.

The first effect of this sulphurous mud-bath is to excite a strong determination of blood to the surface, which becomes so vividly red as to remind the bather of the action of boiling water on a lobster. Soon, however, a profuse perspiration breaks out, which renders it dangerous for the patient to expose himself to the open air for a considerable time after the bath. The therapeutic influence of this application is most evident in chronic articular enlargements, rheumatic arthritis, some indolent tumours, chronic and intractable cases of secondary syphilis, and rheumatism. Few but real invalids are to be met with in Acqui, and even of these there are not many, for the town itself offers little inducement for the mere tourist.

The next Spa in our itinerary is ABANO, the birthplace of Livy, now a mere village, which also lies in the same fertile province of Lombardy as Acqui. Small as the place is, the bath establishment and the accommodation for visitors are both excellent.

The waters issue from the foot of the Euganean hills, off spurs from the northern declivity of the Apennines, and belong to the class of hot sulphurous springs, containing also muriate and sulphate of soda. There are several sources, the hottest of which has a temperature of 181 degrees. Besides the waters, the mud of Abano, like that of Acqui, is used medicinally. This "humus" is taken out of the hot springs, and having been exposed to the air for a long time, so as to bring it down to an endurable temperature, it is used for local and general baths. It is said, that, after the bath, the mud is again returned to the fountain for fresh use. The effects of these mud baths are stimulating and rubefacient, and differ little from those of the baths of Acqui which I have just described.

Tuscany contains several important Spas; the first of which that I visited was PISA. Having, in another work, given a long account of the present state of this once most prosperous of Italian capitals, I could say nothing new on that subject in this place, and shall, therefore, pass at once to the consideration of its mineral waters.

The mineral baths of Pisa have been used medicinally since the middle of the twelfth century, and the present bath establishments are at least a hundred years built. The thermal source rises from a calcareous spar rock, at the foot of Mount St. Julian, where, within an area of seventy paces, there are no less than twelve thermal springs, which vary in temperature from 106 degrees to 81 degrees. They all belong to the class of thermal saline waters, and leave a calcareous incrustation in the wells, and even in the baths; and a pellicle of the same character, consisting of salts of lime and magnesia, floats on the surface of the water, when left undisturbed in the baths for a day and night.

The Pisan mineral springs are used internally in cases of jaundice and other hepatic complaints; in gravel, and some renal complaints, in chlorosis, in dysentery and dyspepsia, attended with pain and vomiting. The warm baths are employed in the treatment of gout and rheumatism, impaired power, and enlargement of the joints; also in certain chronic ulcers and skin diseases, and in some obstinate cases of chronic neuralgia.

From Pisa, a journey of less than an hour by the "Strada Ferrata Livornesi," brings us to LUCCA, five leagues from which are the baths of the same name. They are situated at the foot of Monte Corsena, in one of the prettiest valleys in Tuscany. The watering place itself consists of a long street of handsome hotels, shops, and lodging houses, and nothing seems left undone to render this place one of the most agreeable residences in Italy.

There are five or six bathing establishments, scattered at some distance from each other, in the vicinity of the principal sources, one of which is reserved for the poor gratuitously, and the charges of all are extremely moderate. The supply of water from these sources is very abundant, and they differ little from each other, except in temperature, in which they vary from 88 degrees to 133 degrees. The principal mineral ingredients at the Lucca waters are the sulphates of magnesia, lime, and alumina, together with smaller quantities of the carbonates and chlorides of the same bases, and also silicate of iron, and traces of iodine and bromine.

Some of the bath houses of Lucca have been in use since the sixteenth century, when this place was one of the most fashionable Spas in Europe. And ever since that time the soothing and sedative properties of the mineral waters of Lucca have widely extended their employment in the treatment of cases of chronic rheumatism, leucorrhœa, catarrhal affections of the urinary organs, dyspepsia, and certain cutaneous erruptions.

Besides Pisa and Lucca, Tuscany contains many other mineral springs, as for instance, MONTE CATINO, between Lucca and Pistoia. The two principal sources are both muriated saline waters, having temperature respectively of 68 degrees and 86 degrees. Both these springs contain sulphates of lime and alumina, muriate of soda, and carbonic acid; but in such different proportions, that one source, that known as the "Tettucio," is strongly purgative, while the other, the "Bagnuola" source, is only slightly aperient. The latter is the source generally prescribed, and is used chiefly in chronic enlargements of the liver, over which organ it exercises a powerful influence, somewhat resembling that of the Carlsbad Spa. Monte Catino is also much resorted to in cases of chronic dysentery.

Near Florence are the warm springs of SAN CASCIANO, and in the same department the sulphurous waters of VOLTERRA, with several others. In the Roman States the only springs I had any opportunity of seeing were the warm saline waters of CIVITA VECCHIA, which have a temperature of 86 degrees. The Campagna also abounds in hot sulphurous springs, and at VITERBO and PORRETTA are similar sources. The latter of these is the most important. It is strongly sulphurous and very gaseous, this gas consists principally of carburetted hydrogen, a circumstance of which the guides often take advantage, to startle unscientific visitors, by approaching a torch over the well on which the gas takes fire, and a luminous vapour floats over the water. Besides their sulphurous constituents, the sources of Porretta contain a great quantity of organic matter, and salts of lime and soda, with traces of iodine. The taste is bituminous and nauseous, and the temperature varies from 86 degrees to 100 degrees. The action of Porretta water is strongly purgative and diuretic. It is used internally as well as in baths, in the treatment of enlargements and congestions of the abdominal viscera, and in chronic skin diseases.

From the Roman to the Neapolitan Spas the transition is natural; and with a brief account of the latter watering places I shall close this chapter and my work.

Some years ago I had occasion to reside in Naples, during a considerable part of summer. The season was unusually hot, for even the Neapolitan dog days. Every citizen of the modern Parthenope who could do so, had left town for some cool, maritime retreat; the very lazzaroni seemed prostrated by the long enduring sirocco. And the last stranger, but myself, having fled, I was left to the solitary enjoyment of one of the largest hotels in the Chiaja. I felt conscious that if I could not succeed in finding some occupation to wile away the time which it was necessary I should remain in or about Naples, I should soon fall into the lazzaroni condition of mental torpitude. I had no books, nor could I procure any, except bad Italian translations of worse French novels; I had some acquaintances, but they were all at the sea side. Theatres and operas were out of the question in such an atmosphere; I had seen all the sights over and over, and could not, like some of my Neapolitan friends, sit complacently over an ice in the Café de Paris from morning till night.

At last a kind friend suggested to me that I might with advantage, combined with amusement, visit and examine the various watering places in the neighbourhood of Naples. I gladly embraced his counsel, and shall be well satisfied if the following account of these Spas prove as useful to others, as the labour of acquiring the information here recorded proved to myself.

The most celebrated of the Neapolitan watering

places is CASTELLAMARE, occupying the site of the ancient Stabiæ. The town, which contains a population of about seventeen thousand inhabitants, is built along the sea shore, and is sheltered by Monte Sant' Angelo from the east wind. Facing, as Castellamare does, the west, it thus enjoys a great amount of shade, and is protected from the excessive glare and heat of the mid-day sun, so distressing in summer, on the opposite side of the bay.

The mineral waters of Castellamare enjoy a great reputation with Neapolitan physicians, in the treatment of chronic gout and rheumatism. These springs were analysed some years ago by command of the late government, and were found to differ widely in their composition. Four of these sources contain salts of iron, the principal of the chalybeate springs being the "Acqua Ferrata," rising in the Strada Cantieri. Four of the wells are saline, their chief ingredients being muriate and sulphate of soda, with chloride of calcium; and four are sulphurous and chalybeate, containing sulphate of iron, with a large volume of sulphuretted hydrogen gas. These several springs, though so distinct in character, all rise within a small area at the base of the hill on which the town stands.

The physiological and therapeutic action of these various mineral springs are, as is indicated by the foregoing classification, necessarily very different. The ferruginous springs, especially the "Acqua Ferrata del Pozzillo," are very powerful chalybeates.

The "Acqua Media" is a saline aperient, not unlike Seidlitz water in taste.

The action of the four sulphurous springs of Castellamare need not be here described, as they are similar in effect to other waters of the same class, and are chiefly used by the local physicians in the treatment of chronic skin diseases, and arthritic affections.

Off the opposite side of the Bay of Naples from Castellamare, lies the island of Ischia, the next and last of the watering places which I have now to describe. Ischia, which is the largest island in the Bay of Naples, is about twenty miles distant from the city, and two miles from the isle of Procida. Ischia presents, on every side, the most unmistakable proofs of the vast volcanic convulsions of which it has been the scene, and throughout its whole circumference of twenty miles the traces of violent igneous action are everywhere visible in the physical aspect of the country, which appears chiefly composed of trachytic and lavic rocks. And on the side of one mountain alone, viz., Monte Epomeo, twelve separate volcanic cones may still be traced. This mountain, which gives Ischia its peculiar pyramidal aspect, rises in the centre of the island, in successive series of highly cultivated terraces, to a height of between two and three thousand feet.

Ischia is much frequented by the Neapolitans, as a cool summer residence, but its chief recommendation at all seasons is the reputation of its mineral waters.

There are three towns on the island, in each of which tolerable accommodation may be obtained. Of these towns, or rather villages, CASAMICCIOLA is the most resorted to, as the two others, Ischia and Foria, are exposed to the hot and relaxing south and west winds. Casamicciola, or Casamiccia as it is often called, which is close to the principal mineral sources, is very prettily situated behind Lacco, on a rising ground near the sea; it has a population of about three thousand inhabitants, and contains two or three comfortable boarding houses. In 1859, when I was in Naples, Casamicciola contained a most admirable institution, namely, a sanatorium for convalescent patients discharged from the city hospitals, which was founded in the year 1600 by the *Monte della Misericordia*, one of the religious confraternities of Naples. And I may here observe that there are few charitable institutions of which there is more necessity in this country. The pressure on our hospitals is so great that patients are, perforce, discharged as soon as they are pronounced convalescent, to make way for the admission of more urgent cases. And few but medical men who have to witness the sufferings and privations of these weakened and broken down, poverty stricken, convalescent patients, discharged from hospitals, can adequately realize how great is the necessity for the establishment in the vicinity of all great cities of some convalescent hospital, like that of Ischia. However, as some years have now elapsed since my stay in Naples, it is not improbable that the change of government, which, since then, has led to the suppression of monastic institutions, hospitals, and almshouses, has also long ere this abolished the convalescent religious hospital of Ischia, as a relic of monastic ignorance of the laws of political economy.

The principal mineral sources of Ischia, according to Dr. Chavelly, a resident practitioner, are the " Gurgitello," " Acqua di Citara," " di Cappone," " d'Olmitello," and the " Acqua di Bagno Fresco," all of which springs are thermal, as well as mineral waters. The Gurgitello, which is the most frequented of these springs, rises in the Val Ombrasco, near Casamicciola. The temperature of the water is one hundred and sixty-seven degrees, and its chief ingredients are muriate of soda, carbonate of soda, and sulphates of soda and lime, together with a large amount of free carbonic acid gas. Its medical use is almost entirely in the form of baths, in cases of disease of the periosteum, and in the treatment of sciatica and chronic rheumatism.

The Acqua di Citara rises on a sandy bay, about a mile south of Foria. It varies in temperature from one hundred and fifteen degrees to one hundred and twenty degrees, and its principal constituents are the sulphate and muriate of soda. Its action is chiefly that of a refrigerent and saline laxative, and it is, moreover, used specifically in certain uterine complaints, and in some diseases of the female breast.

The Acqua d'Olmitello contains soluble salts of soda, lime, and magnesia, together with free carbonic acid gas, and is employed in renal and urinary disorders, cutaneous affections, and some hepatic diseases.

The Acqua di Cappone, as its name implies, resembles Wiesbaden water in its similarity of flavour to weak chicken broth. The temperature is ninety-eight degrees, and it differs from the Gurgitello, principally in strength. The chief use, however, as its ancient name "Acqua del Stomacho" implies, is in dyspepsia and chronic gastro-intestinal derangements, and also in some uterine affections.

The "Acqua di Bagno Fresco" is an alkaline water of the same character as the last described spring near which it rises. Like Schlangenbad or Wildbad, and with as much truth, it is said to possess the property of rendering the skin white and soft; I need hardly add that it is therefore specially resorted to by the Neapolitan fair. This source is, moreover, prescribed in some chronic opthalmic affections, as well as in certain cases of chronic skin diseases when a stimulating remedy is required. The Bagno Fresco water is also used as a preparatory Spa before commencing the use of the Gurgitello.

The "Stufe," or natural vapour baths, of Castiglione, in the mound of lava, close to Lacco, are heated by steam, rising through crevices, in two small chambers, in the mass of lava. The temperature of these steam baths is one hundred and thirty degrees, and they are employed in the treatment of cases of chronic rheumatism, gout, and rheumatic arthritis of long standing. They are also sometimes prescribed, but with great caution, in certain cases of partial paralysis. I would not myself, however, sanction their employment in any case of paralysis under my care. The "Stufe" are recommended by the local physicians in cases of scrofula, especially when confined to the external glandular system, and in some obstinate chronic skin diseases. As, however, these baths are powerful stimulants and excitants, they should on no account be used, in any case whatever, without the advice and sanction of a local physician.

My task ends here. I have now accompanied my reader through the principal watering places of continental Europe, and, as I trust, aided the valetudinarian traveller in pursuit of health, with the counsel of his medical adviser, to select the mineral water most suitable for his condition.

I cannot conclude, however, without expressing my obligations to the resident physicians at nearly all the watering places I visited, for the valuable professional information they favoured me with, as well as for the personal attention I received from many of these gentlemen. I have to do so more especially to Dr. Velten, of Aix-la-Chapelle; Dr. Genth, of Schwalbach; Dr. Haussmann, of Wildbad; to the proprietors of Schinznach; Dr. Barthez, of Vichy; and to the late Dr. Smith, of Eaux Bonnes and Pau. I am indeed much indebted to my medical *confrères* abroad, for the opportunities so freely afforded me of studying the effects of mineral waters on disease practically, in the sanitary institutions or hospitals attached to most of the continental Spas. The experience of the "Spa-doctors" has furnished me with many valuable hints, which I made use of, and have duly acknowledged.

I have, however, found myself obliged to differ from the opinions of several of the most eminent writers on this subject on some points: such, for instance, as the question of the use of certain mineral waters in the treatment of chronic pulmonary disease, and especially as to their alleged efficacy in the treatment of phthisis. I should be very reluctant to dissent thus from opinions so generally adopted did I not feel conscious that the views contained in the foregoing pages have not been formed without sufficient experience of the mineral waters I have described, and of their sanative action. For several years I have studied the principal Spas on the continent during journeys undertaken for the special purpose of collecting original information on the subject of this work. I have, moreover, endeavoured to trace the influence of mineral waters on disease, not only at the Spas abroad during "the course" of the waters, but also at home, by comparison of their effects with those of other remedies prescribed in similar cases. Having in this way acquired some experience, and taken pains to gain accurate information, I have recorded the results of these observations, even when they do not coincide with the conclusions arrived at by other observers; for such differences of opinion but prove the wisdom of the first aphorism of the Father of medicine that in our art, as in all others, " Experience is fallacious, and judgment difficult."

Since I last visited the Spas of Germany, a year ago, the great political changes which have taken place in that country have altered the form of government in two or three of the places described in this volume. Since the earlier pages of this work were placed in the printer's hands, such changes have occurred especially in the no longer " Free City " of Frankfort, and the *Brunnens* of Nassau, and Hesse Homburg. My account of these places was written immediately before the late war, and as there is no probability of any change being affected thereby, in the social aspect of the localities of these Spas, excepting the suppression of a few gambling tables, I have nothing to add to that account of them, or alteration to make in it.

Nothing now remains for me to add, in conclusion, but the hope that the physician who may honour this book by a perusal, may find herein information in some respects new, and in all, at least, accurate and serviceable; and that the valetudinarian reader may derive benefit from it. Should this work thus assist either physician or patient, in the pursuit of health, I shall not have written in vain.

THE END.

# INDEX.

Aachen (see Aix la Chapelle), 107—117.

Abano, the baths of, 356—357.

Abstinence, its good effects in some cases of dyspepsia, 38; also in gout, 52, 308.

Acqua Ferrata di Rio, 3.

Acqui, the town and mineral waters of, 355—356.

ADELHEIDSQUELLE, mineral springs of, 12.

Ailments, list of the principal ones treated by mineral waters, ii.

AIX-LA-CHAPELLE, 107. Hotels *ibid*. The town, 109. The sulphurous springs and their remedial effects, 109 to 113. The thermal baths, 113—114. Cases in which their use would be dangerous, 115.

Aix-Les-Bains, 290—291. The thermal establishment, 292. The mineral water caves, 293—294. The thermal sources and their medicinal uses, 295—296.

Alkaline waters, 11.

Alterative medicines, their use before mineral waters, 20.

Amélie-Les-Bains, its climate and thermal mineral waters, 334—335.

Anæmia, treatment of, by chalybeate mineral waters, 7, 104, 140, 178, 199, 227, 305, 344, 362.

Aperient bitter waters, 243.

Apoplexy, excited by hot baths—219, 342, 343.

Apoplexy, treatment of, by saline waters, 243.

Appetite, effect of travelling on the 18.

Austrian officials, their civility, 204.

Argelèz, 335.

Auteuil, and its mineral waters, 349, 350

BADEN-BADEN: the town, 258, 259, 260. The gambling rooms, 261. The mineral waters, their composition, and medicinal uses 262—264.

Baden-on-the-Limmat, 266. The mineral springs, 267 Old writers on Baden quoted, 268—270. Diseases in which this Spa is still resorted to, 271—272.

Baden near Vienna, 8.

Bagnères-de-Bigorre: the place described, 317—318. The sulphurous waters and their remedial effects, 318—320.

Bagnères-de-Luchon: the town, its situation, and resources, 332—333. The springs and baths, their chemical composition, and therapeutic action, 333—334.

Barèges: the village, its wild and romantic position, 330. The sulphurous waters and their effects described, 331.

Bath, the warm springs of, 12, 51.

Baths, general observations on the thermal baths, 22 Their classification and remedial action, 23, 24, 25. The mode of using them, 26. Injurious consequences of their incautious use, 27, 28, 29.

Bavaria, the railways of, 185. The country, 187. The mineral waters, 191, 198, 199. The peasantry, 189—190.

Beer, German, described, 67.

Belgium, 83—86

Bilin, "The German Vichy," 239, 240,

"Bitter Waters," account of the, 11, 240, 243. Their effects and medicinal uses, 12, 243, 244.

## INDEX.

Black Forest, a drive through the, 243–249.
Bocklet, the watering place, 198. The saline springs, 199.
Bohemia, a tour through, described, 204–221.
Borcette (or Burtscheid), 115–117.
Bourbonne-Les-Bains, 10.
Bright's disease, treatment of, by Bilin water, 240.
Bronchitis, chronic, mineral waters used in the treatment of, 125, 297, 326, 335.
Bromated mineral waters, 12.
BRUCKENAU, town of, described, 199. The chalybeate springs and their uses *ibid*.
Brussels, 86.
BUXTON, 51.

Cannstadt, 245 The mineral springs, 246–247. Their medicinal properties, 248.
Carbonic acid, in mineral waters, 2, 3, 133, 305.
Carlsbad: the town, its situation, and resources for visitors, 206–210. The mineral springs anylized and described, *seriatim*, 211–214. The effects of the waters on health and on disease 216–217. The remedial uses of Carlsbad Spa, 217–219.
Castellamare and its mineral waters, 362–363.
Cauterets: our walk to this Spa described, 320–323. The town, thermal establishments, and environs, 324–325. The sulphurous sources and their therapeutic properties, 326–327.
Chalybeate mineral springs, general observations on their composition and remedial action, 6–8.
Chaudfontaine, 87–88. The baths and thermal springs, 89–90.
Cheltenham, 8–10.
"Chemically-Indifferent Springs," 13.
Civita Vecchia, 360.
Club, a German, described, 167.
Col de Tort, our journey across the, 336–337.
Cold bathing, remarks on, 57–58.
Cold mineral waters, observations on, 2–3.

Cologne, 120.
Constipation, habitual, treated by mineral waters, 196–243–305.
Consumption, see Phthisis.
Cookery, German described, 59–127.
Country life: its physiological effects on citizens, iv, v.
Crontal, 8.
Cursaals, the German described, 62, 63, 66, 260, 262.
Cusset, and its alkaline sources, 308–309.
Cutaneous diseases, influence of mineral waters on various forms of, 197, 256, 277, 297, 334, 353, 360, 367.

Dax and its thermal waters, 345–347.
Dietetics, 30. Idiosyncrasies on this subject, 31. Various systems of living, 32. Influence of diet on health and disease, 33. Animal food, its use and abuse 34.
Diabetes, occasional efficacy of Vichy water in cases of, 306.
Diligence: travelling in Germany, 129.
Dyspepsia, its nature, causes, and treatment, 30–36. The influence of mineral waters in cases of dyspepsia, 37–38. Treatment of indigestion by various mineral waters, and cases which are adapted for each Spa, 104, 154, 218, 243, 247, 288, 332, 334, 358, 359.
Dyspepsia, Gouty, waters used for, 50–307.

Earthquake of Lisbon: its effect on some thermal springs, 5.
"Earthy-springs," account of the, 12.
Eaux Bonnes, 337. Sources of Mont-Trésor *ibid*. Their chemical composition and curative properties, 338–339.
Eaux Chaudes, 339. The Valley of Ossau, the village, 340. The thermal establishment and warm sulphurous springs, 340. Their efficacy in various diseases, 341. Cautions respecting their use, 342. Cases in which they should or should not be used, 343.

INDEX. iii.

Egyptian railway travelling, incidents of, 106.
Eilsen, 8.
Electricity in mineral waters, 15.
Elmen, 12
Eltville, on the Rhine, 126—127.
EMS, our journey to, 122. The town of, 124. The mineral springs, 124. Their remedial effects in nervous and dyspeptic complaints, 125.
Empress of the French at the Spas, 131–137.
Enghien-Les-Bains, 350—353.
English Spas used in cases of gout,51.
English travellers, 70—76.
Enz, the valley of the, 248—249.
Exercise: that of travelling, its beneficial effects, 17. Its utility in cases of dyspepsia, 38. In gouty cases, 52, 308.

Fever: the Spa fever described, 153, —217.
Fire, scene at a, 118—119.
Flörsheim, 157.
Food : its influence on health and disease, 30. Author's rules for taking food, 35.
Frankfort-on-the-Maine, 166—170.
Franzensbad Spa, 230. The mineral waters, 231—232. The "mud baths" and their sanative effects, 233–234.
Friedrichshall, bitter springs, 11, 243.

Gout, its nature and causes, 40. Premonitory symptoms, 41. Regular gout described, 42–43. Irregular gout, 44. Treatment, ancient and modern, 45—48. Mineral waters in cases of. Their efficay, 49– 50. Spas resorted to, 50 —51. Condition of the blood in gout, 52. Its cure by abstinence and exercise, *ibid*. Its treatment by various Spas: Wiesbaden, 153, 154; Homburg, 176 ; Kissingen, 197 ; Carlsbad, 218 ; Wildbad, 256 ; Aix-les-Bains, 295,296; Vichy, 307, 308.
Grafenberg wine, 128.
Glandular enlargements, efficacy of Teplitz Spa in their treatment, 238. Ditto, Cannstadt, 247. Ditto, Wildegg, 280.

Gambling, German, described, 62— 63. Bad effects on health of gambling, 64—66. The gambling tables of Baden - Baden, 260— 262.
Gas baths of Marienbad, 229. Of Franzensbad, 232.
Gastein, 1, 13.
Geilnau, 11, 50.
Germany.—General remarks on the German Spas, 53. Travelling in Germany, Hotels, &c., 54–60. The people and their characteristics described 61, 67 69, 157, 162, 166, 168. German ladies, a specimen of, 201.

Hæmorrhage, danger of warm baths in cases of tendency to, 28, 113, 219, 343.
Hæmorrhoids, a German idea concerning, 161.
Hall, 12.
Hampstead water, 1.
Harrowgate, 9.
Haussmann, Dr., of Wildbad, 250.
Hercynian Forest, the, 185.
Hof, the town of, 202.
HOMBURG-ON THE-HILL, 50. The town of, 170. The Cursaal, 171. Geological formation of the country and the mineral springs, 172. The " Elisabethquelle," 173. The " Stahlbrunnen " *ibid*. The "Kaiserbrunnen," 174. The "Ludwigsbrunnen" *ibid*. Manner of taking the Homburg waters, 175. Their physiological effects and utility in cases of gouty dyspepsia, 176. In hypochondriasis, *ibid*. Other diseases in which they are beneficial, 177. Cases in which their use would be dangerous, 178 —179.
Hotels, German, described, 56.
Hypochondriasis benefited by bitter waters, 244. By Carlsbad water, 218. Kissingen, in cases of, 196. Eaux Bonnes, in, 339
Hysteria, its treatment by Spa waters, 332.

Indigestion, see dyspepsia.
Iodated springs, 12, 278, 280.

# INDEX.

Ischia, situation and geological formation of the island, 363. The convalescent hospital, 364. The mineral springs and their uses, 365, 366. The *Stufe*, or natural vapour baths, 366, 367.
Italian Spas, general remarks on the, 354, 355.

Joints, diseases of, benefited by Wildbad baths, 257.
Journey, benefit of a, to invalids, v., 17.

Kidneys, diseases of, treated by mineral waters, 240, 304, 306, 335.
Kissingen, our route to, described, 185—190. The own, 190—191. The mineral springs, 192—195. Diseases in which they are prescribed, 196—198.
Kreuznach, 12.

Lahn, the valley of the, 123.
Lambe, Charles, quoted, iv.
Langenbrücken, 8.
Leuk, 5, 8.
Lippspringe, 12.
Liquids, their use at meal time, 37.
Liver, chronic, diseases of, benefited by mineral waters, 196, 243.
Living, mode of at the German Spas, described, 53, 58, 61, 66, 67.
Lourdes, 321.
Lucan Spa, 8.
Lucca, mineral waters of, 12, 358.
Luggage, advice regarding, 78, 122
Lux, 329.

Marienbad, 222. Hotels and resources for invalid visitors, 223. The mineral wells and their medicinal uses 224, 225, 226, 227, 228. The gas baths, 229—230.
Marlioz, 296. Composition and curative effects of the mineral water, 297.
Meals, rules for valdetudinarians on this subject, 34.
Medicines to be taken at the Spas, 20.
Mental rest: its effects on bodily health, iv.

Mind; influence of travelling on the, 71—74. Supposed action of mineral waters in some mental disorders, 216.
Mineral Waters.—Account of the patients by whom they are most frequented, ii. — iv. Mineral springs, divided into cold and thermal, 2 Their mode of origin and chemical composition, 2—3. Their classification, 5, 6. Their therapeutic effects, 14, 15, 16. Author's experience of their use in various diseases, 14—19. General directions for their use, *ibid*— 21.
Moffat, 9.
Mont Dore Les-Bains, 310—311.
Monte Catino, 359, 360.
Moryson, Fynes, cited 60, 270—271.
Moutiers, 5.
Mud baths of Marienbad, 229. of Franzensbad, 233. Their effects and medical application, 234. The Mud-baths of Acqui, 355, 356. Of Abano 357.
Muriated saline waters, 9, 10, 11.

Nassau, the Spas of, 123, 131, 142, 146, 157, 163. The Ex Duke of, 143. The annexation of to Prussia referred to, 265.
Naters, in Savoy, 5.
Nauheim, drive to, from Homburg, 179. The town and cursaal, 180. The saline springs, 181. Their remedial action, 182.
Nervous Complaints, efficacy of mineral waters in their treatment, 23. Various Spas resorted to in these cases, 145, 256, 288, 332.
Neuralgia and Sciatica treated by thermal waters, *see* nervous complaints and 263.

Oberlahnstein, 121—122.
Obesity, Carlsbad water in the treatment of certain cases of, 217.
Old age: waters recommended for the old, 45.
Ostend, 85.
Organic diseases, danger of thermal or stimulant waters in cases of, 22, 218.

INDEX.   V.

Pain, allayed by sedative thermal waters, 24.
Paralysis, mineral waters in some forms of, 238, 256.
Paris, mineral springs in the vicinity of, 349-353.
Passy, chalybeate spa, 349.
Pau: the "Source de Parc," 344.
Pedestrian travelling, author's reminiscences of, 81, 82, 315. Its influence on health, *ibid.*
Pfeffers, 283—284. The thermal sources, 285. History of Pfeffers, 186. Analysis of the waters, 287. Cases in which they are used, 288.
Phthisis: remarks on the employment of several mineral waters in the treatment of this disease, 103, 125, 297, 326, 338, 339.
Pisa: mineral waters, 12, 357 358.
Porretta, 360.
Poverty of blood, *see* Anæmia, also 217, 332, 319, 349.
Püllna, the "bitter water" of 240—241.
Pulmonary diseases, chronic, Spa waters in treatment of some forms of, 167, 297.
Pyrenean Spas, general observations on the, 316
Pyrenees, author's pedestrian journey through the, 313-343. The scenery of the High Pyrenees, 315, 317, 322.
Pyrmont, 7.

Ragatz, the baths of, 282.
Railway travelling in various countries, 54—56, 105—107.
Religious observances of foreign countries, respect due to the, 77—78.
Respiration, effects of travelling on the, 18.
Rheumatism, chronic, and its treatment by thermal waters, 23, 113, 155—156, 256—257, 263, 277, 295, 331, 341, 348, 356, 357, 358.
Rhine, reminiscences of the, 120. 129.
Roman Baths, remains of, at Baden-on-the-Limmat, 267; at Aix-Les-Bains described, 296.
Royat, in Auvergne, 312.

Saidschütz, 242—243.

Saline mineral waters: their classification and medicinal uses, 10—11.
Sea-voyages, their effect on health, &c., 84.
Sedative action of certain Spas, 23, 332.
Sedlitz mineral water, 241—242.
Schinznach, 273. The bathing establishment, 274. The sulphurous water and its effects, 275—276. Diseases in which this Spa is resorted to, 277—278.
Schlangenbad, 241—143. The waters and their uses, 144—145.
Schönau, 236.
Schwalbach, *Langen*, 131—132. The sources, 133-135. The baths, 136—137. Physiological action of these waters, 138. Their composition, 139. Diseases in which they are useful or otherwise, 140—141.
Schwartzenberg, 203—204.
Schweinfurth, 188—189
Sciatica, treated by Wildbad water, 256.
Scrofulous diseases, mineral springs beneficial in cases of, 197, 277, 335.
Skin diseases, chronic, their treatment by various mineral waters, 24, 256, 276, 334.
Smoking, its effects on health, 67—69.
Soden, the town and its saline springs, 163-164. Cases in which resorted to, 165.
Spa, in Belgium, 91—93. The mineral sources, 94—102. Their remedial properties and uses, 103-104.
Spa-Life, routine of, described, 66—67. An incident in, 74—76.
Spas, English and foreign, contrasted, 53.
Spleen, mineral waters in chronic enlargements of the, 196.
Sprudel, of Carlsbad, 211. Of Cannstadt, 248.
Sterility, mineral waters used in some cases of, 104, 197.
Stimulants, their use and abuse as general articles of food, 35—36.
St. Nectaire, 311.
St. Sauveur, 331. The springs and their uses, 332.

Sulphurous mineral springs: general remarks on their chemical composition and physiological action, 8—9. Examples of, 275, 333, 339, 351, 352.
Sutro, Dr., quoted, 139, 164.
Swiss Baden, account of the, 266—272.
Switzerland, a journey through some parts of, 266—289.

Table d'Hôtes, the German, described, 59, 62, 108.
Tamina, the river, 282—285.
Tarasp, mineral water, 10.
Teeth, early decay of the, in Germany, 61.
"Telluric-heat," so called, in mineral springs, 15.
Teplitz, its situation and environs, 235. The town, 236. Number of thermal sources, 236. Their chemical composition, 237. Their general action, 238. Are chiefly used for bathing in, *ibid*. Cases for which they are suitable or unsuitable, 238—239.
Thermal baths, their remedial influence, 23, 24, 25. How they should be taken, 26. Danger of their incautious use, 27—29.
Thermal Waters, their origin, 3.
Tourists, specimens of, described, 77.
Travel, the art of, 70—82.
Travelling, its beneficial effects in various chronic complaints, 17, 18. German modes of, described, 54. Uses and advantages of travelling, 71—72. Companions' choice of, 80, 249—250. Travelling on foot, its enjoyments, 81, 82, 314, 315.
Travellers, English abroad described, 72—74.
Tracts, religious, a present of, 148.
Tüffer Spa, 13.
Tuscany, the Spas of, 357.
Tunbridge Wells, 8, 51.

Valencia, a scene in the Custom House of, described, 79—80.

Velten, Dr., of Aix-la-Chapelle, 108.
Vichy, 11, 50. Route to Vichy, 297. Vichy-la-Ville and Vichy-les-Bains described, 298. Population and visitors, 299. The Emperor's visits and their results, *ibid*. The thermal establishment, 300. Geological formation of the country about Vichy, 301. The mineral springs, the analysis, situation, dose, and medical effect of each source separately described, 301, 302, 303, 304, 305. General observations on the effect of the Vichy waters, 305. Their action in cases of diabetes, 306. In gout, 307—308.
Vineyards, the, of various countries contrasted, 185—187.
Viterbo, 360.
Volcanoes, their connexion with thermal waters, 3—5.
Volterra, Spa of, 360.

Wallenstadt, Lake, 281.
War of 1866, changes occasioned in German watering places by the, 265, 369.
Warm baths, natural, 22. Their medicinal efficacy, 23—24. Directions for their use, 26. May bring on hæmorrhage or apoplexy, 28, 334, 343. Increase obesity, 29.
Watroz, mineral spring, 103.
Weilbach, the bathing establishment, 158. The mineral water, 159. Mode of living at Weilbach, 160. Medicinal action of the water, 161.
Wiesbaden, the capital of the Ex-Duchy described, 146—149. The Kochbrunnen and other springs, and their analysis, 150—151. Physiological effect of the water, 152. Its use in cases of gout and dyspepsia, 153—154. The mineral water baths and their uses, 155—156.
Wildbad, situation of, 248. Journey to, 249. Hotels, 250. The town, 251. Grand bath house, *ibid*. The baths, 252. Hospital, 253. Curious privilege, *ibid*. Geological formation, *ibid*. Com-

position of the waters, 254. Their physiological action, 255. Mode of using the baths, and diseases in which they are employed, 256—257.

Wildegg, 278. The iodated water, 279. Cases in which this Spa is resorted to, 280.

Wine, its use and abuse, 35. German wines, 188.

Würtemberg, the mineral springs of, 245. Attention paid by Government to the improvement of the watering places in, 251.

Würzburg, the city of, 188

Zurich, the Lake of, 280—281.

CPSIA information can be obtained
at www.ICGtesting.com
Printed in the USA
BVHW041630240122
627018BV00010B/347